Official Methods for
the Determination of *Trans* Fat

Second Edition

Official Methods for the Determination of *Trans* Fat

Second Edition

Magdi M. Mossoba
Food and Drug Administration
Center for Food Safety and Applied Nutrition
Office of Regulatory Science
College Park, Maryland

John K.G. Kramer
Guelph Food Research Center
Agriculture and Agri-Food Canada
Guelph, Ontario, Canada

AOCS
PRESS

Urbana, Illinois

AOCS Mission Statement

To be a global forum to promote the exchange of ideas, information, and experience, to enhance personal excellence, and to provide high standards of quality among those with a professional interest in the science and technology of fats, oils, surfactants, and related materials.

AOCS Press, Urbana, IL 61802

ISBN-13: 978-1-893997-72-1

Library of Congress Cataloging-in-Publication Data

Official methods for the determination of trans fat / Magdi M. Mossoba
... [et al.]
  p. cm.
Includes bibliographical references
  ISBN 1-893997-45-6 (obk. : acid-free paper)
  1. Food--Fat content. 2. Trans fatty acids. 3. Gas chromatography
  4. Spectrum analysis. I. Mossoba, Magdi M.

  TX553.L5034 2003
  664′.07--dc21

2003009521

Printed in the United States of America.
13 12 11 10 09 6 5 4 3 1

The paper used in this book is acid-free and falls within the guidelines established to ensure permanence and durability.

Contents

Preface

The authors of this monograph have described the most common gas chromatographic and infrared spectroscopic official methods required for the determination of *trans* fatty acids for food labeling purposes. They have included a brief introduction describing the general structure of fatty acids and the occurrence of *trans* fatty acids in different matrices. They compared the distribution profiles of *trans* 18:1 positional isomers of natural origin that result from biohydrogenation in the rumen, of man-made partially hydrogenated vegetable oils, and of human milk. The consumption, biochemistry, and health effects of *trans* fatty acids have been the subject of numerous studies and critical reviews in the past several decades and are beyond the scope of this monograph. However, the authors have noted the recent advances in conjugated linoleic acid (CLA) research investigating potential anticarcinogenic properties and other beneficial physiological effects in animal studies attributed to *cis,trans*-18:2 isomers of CLA.

In this monograph, the authors have reviewed the current status and limitations of the latest official methods used for the determination of *trans* fatty acids. They have included discussions on the numerous factors that may have an impact on accuracy and precision, including the nature of the food matrix being analyzed, the methodology and experimental conditions used, and the skills of the analyst performing the quantitation of *trans* fats. *Trans* fatty acid methods are described that require the use of long, highly polar gas chromatography (GC) capillary columns, as well as those that apply transmission or attenuated total reflection (ATR) infrared spectroscopy, including the negative second derivative methodology. The authors also point out issues that are not addressed by any GC official method. These include the elimination of GC peak overlap that requires prior separation of the *cis* and *trans* 18:1 geometric isomers by silver ion-TLC, HPLC or SPE. A critique of infrared methodologies is also highlighted. Outstanding issues that have yet to be addressed by the scientific community include reconciliation of differences in total *trans* fatty acid content that are obtained by chromatographic and spectroscopic techniques.

It is becoming apparent there are several issues that need to be considered. One is a need to simplify the current total *trans* determination. If labeling requirement is simply about the determination of total *trans* with exclusion of CLA, then the current infrared spectroscopic methods would be more rapid and convenient. However, it is becoming apparent that there is a need to be able chromatographically separate individual *trans*-containing fatty acid and CLA isomers, since these fatty acids are known to have uniquely different physiological and health effects. Such methods would be positioned to respond to potential changes in future regulations that would take into consideration the health effects of specific *trans* fatty acid isomers.

The extensive discussion in a monograph dedicated to the determination of *trans* fatty acids may guide and inform current researchers and analysts in nutrition and food chemistry and technology, and should provide a stimulus for pursuing more research and improving *trans* fatty acid methodologies.

About the Authors

In 1980, **Dr. Magdi M. Mossoba** received his Ph.D. in Chemistry from Georgetown University (Washington, DC). From 1980 to 1983 he was awarded a postdoctoral research Fogarty fellowship at NIH/NCI (Bethesda, MD), and in 1983–84 he joined the University of Maryland Cancer Center (Baltimore, MD). Since 1984 Dr. Mossoba has been working for the US Food and Drug Administration (FDA) Center for Food Safety and Applied Nutrition (CFSAN) (Washington, DC; since 2001 College Park, MD) as a Research Chemist. His research interests centered on applying Fourier transform infrared (FTIR) spectroscopy to the analysis of food constituents, contaminants, and additives. His work included the analysis of fats and oils by using GC-MI-FTIR, GC-DD-FTIR, and ATR-FTIR techniques, and has led to the development of infrared official methods for the quantitation of *trans* fatty acids in 1999 and 2000. He is currently validating in an international collaborative study a novel negative second derivative ATR-FTIR methodology for the rapid determination of total isolated *trans* fat. His recent research activities also include the identification of foodborne bacteria by analyzing bacterial cellular fat, and applying DNA microarray technology, infrared microspectroscopy and multivariate statistical analysis. Dr. Mossoba has authored 95 peer-reviewed research papers and many review articles and book chapters. He has served as editor/co-editor for four books and has co-authored two monographs on *trans* fat analysis. In April 2003 he received the FDA's Excellence in Laboratory Science Award and in 2008 the AOCS Dutton Award. For more than a decade, he has been active at AOCS and served as chair of the Analytical Division and as vice-chair of the Uniform Methods committee. He is currently the AOCS Chair of the Books and Special publication committee, and co-Chair of the FDA CFSAN Research Scientists Peer Review Panel.

Dr. John K.G. Kramer graduated from the University of Manitoba (1963 BSc Hons, 1965 MSc) and Minnesota (1968 PhD) in biochemistry and organic chemistry. He was a Hormel Fellow at the Hormel Institute, Austin, MN (1968–70) and an NRC Fellow at the University of Ottawa (1970–71), before joining Agriculture Canada in Ottawa as a Research Scientist (1970–present; relocated to Guelph ON in 1997). He was on sabbatical at the US FDA in Washington, DC and the USDA in Wyndmoor, PA. He is adjunct professor at the University of Guelph (Ontario) and Nanchang (China). Awards include Government of Canada Merit Award (1983), CSP Canola Research Award (1984), and Dutton Research Award (1999). He is Associate Editor of LIPIDS (1988-present), and served on NRC expert committee of Fats and Oils, National Standards Board of Canada, Canola Council of Canada, and CODEX committee of Fats and Oils (1984–98). From 1979–85 he was a core member of Agriculture Canada's group to evaluate the safety of canola oil for human use that culminated in successfully obtaining GRAS status for canola oil in the US in 1985. Publications include over 200 refereed papers, 25 chapters and co-editor on 5 books. He has identified, characterized and synthesized the structure of numerous foods, bacterial, and biological components, and evaluated their nutritional and or toxicological properties. The components include lipids, carbohydrates, sphingomyelins, tocopherols, sterols, and hydrocarbons. He has developed several new methods for their isolation and identification. He is experienced in chromatographic (TLC, GC, HPLC, Iatroscan), spectroscopic (FTIR, NMR, MS), and chemical techniques. In the last few years he has been working extensively on the analysis and characterization of fats from ruminants with special emphasis of *trans* and conjugated linoleic acids. He is a core member that initiated the book series on 'Advances in Conjugated Linoleic Acid Research'.

Official Methods for the Determination of *Trans* Fats
Gas Chromatography and Infrared Spectroscopy

Introduction

The food labeling regulations in Canada (Regulations, 2003), the US (DHHS, FDA, 2003), and other countries (Ratnayake and Zehaluk, 2005) require the mandatory declaration of the amounts of *cis*-monounsaturated and *cis*-polyunsaturated fatty acids, total saturated fatty acids and *trans* fatty acids. On the other hand, the amounts of total n-6 and n-3 polyunsaturated fatty acids may be declared on the food label on a voluntary basis. Unlike total fat, individual fatty acids should be expressed as triacylglycerol (TAG) equivalents.

In this monograph, the latest approved gas chromatographic and infrared spectroscopic official methods used to determine *trans* fatty acids will be reviewed. In addition, a brief discussion is included about analytical procedures that have also been applied to the quantitation of *trans* fatty acids in food products. Further studies are still needed to address outstanding issues, namely differences in the total *trans* fatty acid content obtained by gas chromatographic and infrared spectroscopic techniques, the applicability of methods to determine the *trans* content in different food matrices, and the development of methods to chromatographically resolve the *trans* fatty acid isomeric mixtures to the extent that would be sufficient to permit the exclusion of those with potentially beneficial physiological effects from the calculated total *trans* fatty acid content.

Structure and Occurrence of *Trans* Fatty Acids

The principle components in all lipids from plant or animal sources are the fatty acids that differ mainly in chain length and degree of unsaturation. The double bonds naturally occur in the *cis* configuration and are at unique positions along the fatty acid chain; if two or more *cis* double bonds are present in a fatty acid, and are separated by a methylene group, they are referred to as all *cis* methylene-interrupted fatty acids (Fig. 1). The position of the double bond is defined either relative to the carboxyl group carbon atom (at C1) according to the systematic delta "Δ" nomenclature, or relative to the terminal end methyl group of the molecule when employing the common "n" nomenclature. Thus, oleic acid, which contains a single *cis* double between C9 and C10 of the C18 hydrocarbon chain, is designated as *cis* 9-18:1 (*c*9-18:1) or 18:1n-9. Note that the delta "Δ" symbol is usually omitted. In oleic acid the lone double bond is at C9 from either end of the molecule. The latter "n" nomenclature is preferred for the designation of polyunsaturated fatty acids (PUFA). It is noted that the "n" nomenclature, rather than the synonymous "omega" notation, is currently accepted in the scientific literature. According to the "n" nomenclature, naturally occurring PUFA are designated by the chain length followed by a colon and the number of double bonds in the molecule, and the "n" is followed by the position of first double bond starting from the methyl end of the molecule. In the most common PUFA, the first double bond naturally occurs either at the 3rd (i.e., 18:3n-3 or *c*9,*c*12,*c*15-18:3), 6th (i.e., 18:2n-6 or *c*9,*c*12-18:2) or 9th (i.e., 20:3n-9 or *c*5,*c*8,*c*11-20:3) position from the methyl end of the molecule. Positional isomers are fatty acid having the same chain length but differ in the site of unsaturation between different pairs of adjacent carbon atoms along the hydrocarbon chain, while geometric isomers are fatty acids with the same chain length and position of double bonds, but differ in the configuration about the double bond.

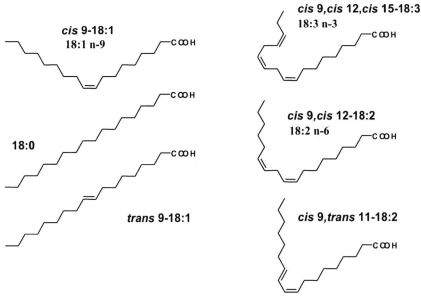

Fig. 1. Chemical structures for common fatty acids in oils and fats.

Trans fatty acids are carboxylic acids that contain at least one double bond in the *trans* configuration. The *trans* containing fatty acids are generally designated using the systematic "Δ" nomenclature. For example, elaidic and vaccenic acids are designated as *trans* 9-18:1 (*t*9-18:1) and *t*11-18:1, respectively. In addition, there are *trans* containing PUFA such as *c*9,*t*12-18:2 or *t*9,*c*12,*c*15-18:3, and conjugated fatty acids that may contain one or two *trans* double bond per fatty acid moiety (Fig. 1). Conjugated fatty acids are defined as fatty acids in which any two adjacent double bonds in a molecule are separated only by a single carbon-carbon bond (instead of one or more methylene groups), such as *c*9,*t*11-18:2 or *c*9,*t*11,*c*15-18:3. The occurrence of conjugated fatty acids will be discussed in greater detail below.

There are two main sources of *trans* fatty acids in the human diet, those produced during processes involving heat and hydrogenation, and those occurring naturally. The major source of *trans* fatty acids in the first group are those produced industrially by partial hydrogenation to harden vegetable oils, while minor amounts are derived from fully refined and deodorized vegetable oils, and any frying process of foods. The second source of *trans* fatty acids are those naturally produced in ruminants, and those present in some plant fats. These will be discussed in turn.

Industrially produced trans *fatty acids*

The major source of *trans* fatty acids in our diet for the past few decades has been those produced industrially during partial hydrogenation of vegetable oils (Ackman and Mag, 1998; Craig-Schmidt 1998; Wolff et al. 2000). Levels of *trans* fat of up to 50% (as percent of total fat) have been reported in products containing partially hydrogenated vegetable oils (PHVO) (Fritsche and Steinhart 1997; Table 1). Similar levels were still found in some products prior to the date when the new *trans* fatty acid regulation came into effect in the United States (Satchithanandam et al. 2004) and Canada (Ratnayake et al. 2007). The *trans*-18:1 isomers in PHVO generally show a random bell-shape distribution from *t*6-*t*8- to *t*13-/*t*14-18:1 (Molkentin and Precht 1995; Ratnayake et al. 2002 & 2006; Cruz-Hernandez et al. 2004). Figure 2 shows the *trans*-18:1 isomer distributions for three different partially hydrogenated oils having a total *trans* content of about 33% produced by different processors. The content of *trans*-containing dienes and trienes depends on whether the oils being partially hydrogenated contained linoleic acid and linolenic acid, respectively, and the extent of hydrogenation. The more partially hydrogenated oils contained correspondingly less *trans*-containing PUFA.

Table 1. Ranges of *trans* Fat Content for Selected Food Products (Fritsche & Steinhart, 1997)

Product	*Trans* 18:1 (% of total fat)	Total *trans* (% of total fat)
Stick margarine, soy	19-41	19-49
Tub margarine, soy	9-21	11-28
Shortening, soy	9-27	3-30
Cooking oil, soy	5-11	1-13
Salad oil, soy	0-3	0-5
Cookies	3-32	4-36
Cake	9-11	10-13
Milk shake	2-3	2-4
Hamburger	3-5	3-5
Potato chips	0-34	0-40
French fries	3-32	3-34
Butter	2-6	2-7
Whole milk	2-3	2-4
Beef	2-5	2-5

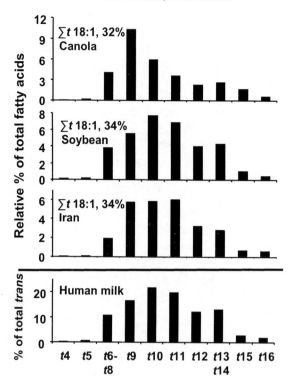

Fig. 2. Relative distribution of *trans* 18:1 positional isomers in different PHVO and human milk. Partially hydrogenated canola and soybean oils are from Canada, and a PHVO of an unspecified oil is from Iran (Kramer unpublished data). The total *trans*-18:1 content in the different PHVO is also provided.

The deodorization step during refining of vegetable oils was shown to produce up to 3% *trans* (as percent of total fat) due mainly to geometric isomerization of linoleic and linolenic acids at temperatures above 200°C (Ackman et al. 1974; Wolff 1992; Buchgraber & Ulberth 2002). Deep frying likewise produced isomerization of linoleic and linolenic acids at temperatures in access of 200°C (Sebedio et al. 1996). Heating vegetable oils at elevated temperatures resulted in 13 to 14 times more isomerization of linolenic acid to $c9,c12,t15$-18:3 and $t9,c12,c15$-18:3 with minor amounts of $c9,t12,c15$-18:3, and $c9,t12,t15$-18:3, than linoleic acid to $t9,t12$-18:2, and $c9,t12$-18:2 and $t9,c12$-18:2 (Wolff 1992; Sebedio et al. 1996).

Generally, the discussion of *trans* fatty acid isomers is limited to 18:1 isomers, which is understandable, since the *trans* fatty acids arising during partial hydrogenation and deodorization of vegetable oils occur as the result of geometric and positional isomerization of corresponding PUFA having the same chain length. Vegetable oils contain mainly C18 PUFA and only small amounts of C16 or C20 PUFA that is reflected in the relative abundance of the *trans* fatty acids. In contrast to *trans* fatty acids derived from vegetable oils, ruminant fats contain more 16:1 and 20:1 *trans* isomers, because these isomers can be derived from biohydrogenation of their corresponding PUFA sources in the diet, as well as by chain-shortening to 16:1 and chain-elongation to 20:1 of their 18:1 precursors (Kramer et al. 2008). Ruminant fats also contain branch-chain fatty acids that coelute by GC in the same region as the mono-unsaturated *cis* and *trans* fatty acids, which complicates the analysis (Precht & Molkentin, 2000a; Kramer et al. 2008).

Trans *fatty acids derived from ruminants*

The other source of *trans* fatty acids are dairy and meat products from ruminants. Rumen bacteria metabolize (biohydrogenate) dietary linoleic and linolenic acids to conjugated and *trans*-containing mono- and poly-unsaturated fatty acids (Griinari and Bauman, 1999; Bauman and Griinari, 2003; Wallace et al., 2007). The *trans* monounsaturated, vaccenic acid (*t*11-18:1), and the *trans* containing conjugated fatty acids, rumenic acid (*c*9,*t*11-18:2), are of special interest because of reported beneficial physiological effects (Belury, 2002). The total *trans* mono-unsaturated fatty acid content in milk lipids may reach up to 23% as percent of total milk fat depending on the diets fed to dairy cows (Cruz-Hernandez et al. 2007), and they includes all possible *trans* isomers of 16:1, 18:1, 20:1, 22:1 and 24:1 (Precht and Molkentin 1996, 2000a, 2000b; Wolff et al. 1998; Kramer et al. 2001; Cruz-Hernandez et al. 2004, 2006). Figure 3 shows the effect of diet on the *trans*-18:1 isomer distribution in milk fat. Vaccenic acid (*t*11-18:1) predominates in milk fat when cows are strictly pasture-fed (Precht and Molkentin 1997; Kraft et al. 2003), while increasing amounts of *t*10-18:1 are evident in the milk fat of cows fed concentrates rich in digestible carbohydrates, oil- or oilseeds rich in PUFA (Piperova et al. 2000; Shingfield et al. 2005; Roy et al. 2006; Cruz-Hernandez et al. 2007), and ionophore antibiotic, such as monensin (Eifert et al. 2006; Cruz-Hernandez et al. 2006). Levels of up to 7% of total *trans* mono-unsaturated fatty acids are also reported in the red meats of ruminants depending on the diet fed (Hristov et al. 2005; Leheska et al. 2008; Aldai et al. 2009). Figure 4 shows the *trans*-18:1 isomeric distribution in the subcutaneous fat of selected striploin steaks purchased at retail stores in Canada (Aldai et al. 2009). Only the beef fat from grass-fed animals showed vaccenic acid (*t*11-18:1) as the predominant *trans*-18:1 isomer (Fig. 4, upper graph), the remaining beef fat samples contained increasing amounts of the *t*10-18:1 isomer. The maximum level of total *trans*-18:1 found in the retail beef sampled was 13%.

Milk and meat fats of ruminants are the major sources of conjugated linoleic acids (CLA) in which the rumenic acid (*c*9,*t*11-18:2) isomer predominates (Parodi 2003). Other CLA isomers include *t*7,*c*9-18:2 (Yurawecz et al., 1998) that is generally associated with dairy cows fed high concentrate diets (Piperova et al. 2000; Aldai et al. 2008), and *t*11,*c*13-18:2 with pasture-fed cows (Kraft et al. 2003). The content of total CLA in ruminant fat can range up to 1.89% (as percent of total fat) in commercial dairy herds (Precht and Molkentin, 2000b), up to 2.87% in

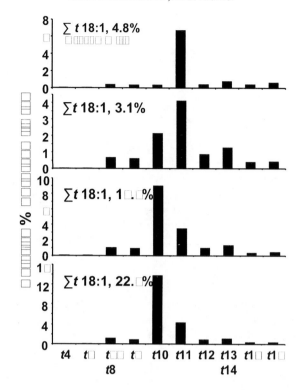

Fig. 3. Relative distribution of *trans* 18:1 positional isomers from selected cows' milk taken from a previous study (Cruz-Hernandez et al., 2007). The total *trans*-18:1 content in the milk fat is provided in the figure.

Alpine milk (Collomb et al. 2001; Kraft et al. 2003), and to about 4% in short term feeding trial of sunflower or safflower oils (Bauman et al. 2000; Bell et al. 2006). The total CLA content in beef was shown to be higher in subcutaneous than in intramuscular fat (Fritsche et al., 2001; Dannenberger et al., 2004 & 2005; Kraft et al., 2008; Aldai et al., 2009), and least in the phospholipid fraction (Fritsche et al., 2001; Nuernberg et al., 2005). The relative proportion of rumenic acid ranged from slightly above 80% in lambs (Nuernberg et al., 2005), to 65 to 78% in pasture- and concentrate-fed cattle (Fritsche et al., 2001; Dannenberger et al., 2004 & 2005; Kraft et al., 2008), and about 60% in the retail beef in Canada (Aldai et al., 2009). The remaining CLA isomers were mainly *t7,c9*-CLA and *trans,trans*-CLA (Fritsche et al., 2001; Dannenberger et al., 2004 & 2005; Kraft et al., 2008; Aldai et al., 2009), none of which have demonstrated beneficial health effects.

CLA isomers have been reported to protect against cancer and show other benefits in animal studies, but the effects are isomer specific (Pariza et al., 2001; Belury 2002; Martin & Valeille 2002). A few epidemiological studies have shown that dairy foods rich in CLA may protect against breast cancer in postmenopausal women (Aro et al. 2000) and reduce the risk of colorectal cancer (Larsson et al. 2005). The authors suggest that rumenic acid ($c9,t11$-18:2), and possibly also its precursor vaccenic acids ($t11$-18:1), are responsible for these effects. It has also been shown that rumenic acid decreases LDL:HDL cholesterol ratio and total:HDL cholesterol, while $t10,c12$-18:2 has the opposite effect on the blood lipid profile in healthy humans (Tricon et al. 2004). In fact, the $t10, c12$ isomer of CLA was shown to act as a cancer promoter in colon carcinogenesis (Rajakangas et al., 2003).

The current rules for mandatory declaration of total *trans* fatty acids in the US (DHHS, FDA, 2003) and Canada (Regulations, 2003) excludes CLA, despite the fact that most CLA isomers contain at least one *trans* double bond in the molecule. The decision was based on the fact that CLA showed no negative responses in the lipoprotein pattern in humans (DHHS, FDA, 2003). However, more recent studies have show that the different CLA isomers have uniquely different biological responses in mammalian systems including humans (Pariza et al., 2001; Belury 2002; Martin & Valeille 2003), and some isomers have proven negative effects (Tricon et al. 2004; Rajakangas et al., 2003). An exclusion of all CLA isomers may only be justified if ruminant fats consists mainly of rumenic acid, which apparently is not the case in concentrate-fed ruminants. Currently intensive feeding practices are common in beef and dairy production which has led to increased levels of *trans* fatty acids and *trans* conjugated fatty acids other than vaccenic and rumenic acids in milk (Roy et al., 2006; Cruz-Hernandez et al., 2007) and meat of ruminants (Kraft et al., 2008; Leheska et al., 2008; Aldai et al., 2009). Therefore, the exclusion of all CLA isomers from mandatory labeling of *trans* fatty acids may need to be reconsidered, if warranted by more scientific research findings on the physiological effects of individuals CLA isomers.

Trans *fatty acids in humans*

The *trans* 18:1 isomers represent over 80% and 90% of total *trans* fats in ruminant fats and partially hydrogenated vegetable oils, respectively (Wolff and Precht, 2002). Consumption of these *trans*-containing food products results in the accumulation of *trans* fatty acids in tissue lipids as well as human milk (Craig-Schmidt 1998; Aitchison et al. 1977; Chen et al., 1995; Mosley et al., 2005). The distribution of the *trans*-18:1 isomers in human milk reflects the relative intake of dairy fats or PHVO in the mother's diet (Wolff, 1995; Molkentin and Precht 1995; Precht and Molkentin 1999 & 2000c; Li et al. 2009). The lower graph of Fig. 2 shows the *trans*-18:1 profile of a human milk fat from the Washington D.C. area, which resembled that of partially hydrogenated soybean oil. This result is consistent with data from a recent study showing that the *trans*-18:1 isomers distribution of human milk fat reflected the diets consumed in five different regions of China (Li et al. 2009).

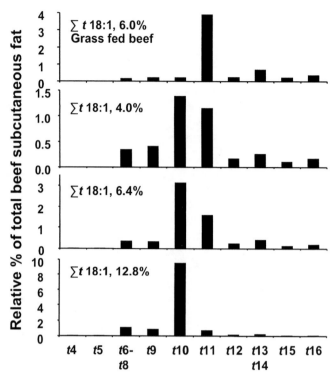

Fig. 4. Relative distribution of *trans* 18:1 positional isomers in different beef steaks sampled in Calgary, AB, Canada (Aldai et al., 2009). The total *trans*-18:1 content in the different beef fats is provided in the figure.

Capillary Gas Chromatography

Gas chromatography (GC) is the most widely used analytical method for the determination of the fatty acid composition in any food matrix. The availability of very long flexible fused silica capillary columns coated with highly polar stationary phases makes it possible to obtain the complete fatty acid profile of a fat product and to maximize the resolution of most *trans* fatty acids. Improvements in GC technology related to reproducibility of injections and accuracy of integrations have greatly contributed to the success of this analytical technique. The chemical composition of the matrices to be analyzed may require product specific derivatization procedures and GC conditions. Over the past 40 years official methods have continually been ungraded to incorporate the ever changing improvements in GC instrumentation and column technology to provide more accurate fatty acid compositions of fat and oil products. The attempt has always been to propose methods based on a single capillary GC procedure. In the following section several of the derivatization techniques and GC operating conditions will be evaluated for product-specific methodologies, to maximize the resolution of as many fatty acids as possible, and to discuss their limitations.

Selection of fatty acid derivative

The most common derivatives used to analyze the long-chain alkyl groups of the different lipid classes are methoxy derivatives prepared by reaction with excess methanol in the presence of catalytic amounts of acid or base. Methoxy derivatives are preferred because of their great volatility and superior resolution by GC (Ulberth et al. 1999). The methoxy products expected from *O*-acyl and *N*-acyl side chains are fatty acid methyl esters (FAME), while alk-1-enyl ethers yield dimethylacetals (DMA), see section and figures below. A wealth of information is currently available for the analysis of many different fat and oil products. The choice of catalyst to prepare methoxy derivatives of different food products will be discussed separately in the next section. Milk and ruminant fats require additional attention because of the presence of short-chain fatty acids from 4:0 to 8:0 that produce volatile FAME. To avoid losses of the short-chain FAME during analysis, Wolff and Fabien (1989) suggested the use of isopropyl instead of methyl esters for dairy fats that have the added advantage of requiring almost no correction for loss in FID response (Wolff et al., 1995). However, this resulted in an increase of the elution time for the long-chain fatty acid isopropyl esters and a poorer resolution of the *trans*-18:1 isomers. Analyses of both methyl and isopropyl esters were occasionally performed and the results were subsequently combined; see Piperova et al. (2000) as an example. Unfortunately, details of how these data sets were combining was never described, and besides, this would not satisfy the requirements of an official method that seeks to find a single GC separation. An alternative method was developed by Christie (1982) and that was subsequently modified for milk fats by Chouinard (1999). It involved the preparation of FAME in an aqueous-free system using Na methoxide in GC autosampler vials filled nearly to the top to minimize the head space and loss of volatile FAME (Cruz-Hernandez et al., 2004). After the reaction Na ions were precipitated using oxalic acid, and the upper phase was analyzed directly by GC. To remove the precipitate and clarify the hexane, the vial was placed for about 10 min into a freezer at -20°C, then centrifuged, and the upper layer was passed through a small column (pasture pipette) with a little anhydrous Na_2SO_4 before analysis by GC (Cruz-Hernandez et al., 2004 & 2006). The use of test tube as a reaction vessel to methylate 20 mg of dairy fats, and then cooling the mixture before extraction, may not be enough to quantitate short-chain FAME (Mendis et al., 2008). Unfortunately, these authors did not report the content of the volatile FAME.

Selection of a catalyst for derivatization of lipids

The selection of a catalyst requires an understanding of the chemical nature of the product to be analyzed, since the lipids may include neutral lipids, phospholipids, glycolipids, sphingolipids and FFA (Kramer and Zhou, 2001; Cruz-Hernandez et al. 2004 & 2006). The catalyst should quantitatively convert all acyl moieties to FAME, not react with any other functional group in the acyl moiety, or ignore certain lipids.

Acid-catalyzed methylations are applicable to the methylation of all common lipid structures, including FFA, *O*-acyl (esters, glycosides), *N*-acyl (sphingomyelin), and alk-1-enyl ethers (plasmalogenic lipids), except ethers (Table 2). FAME are the product of all *O*- and *N*-acyl lipids, while DMA are produced from the alk-1-enyl ethers. A clean 5% (w/v) solution of methanolic HCl can be easily prepared by bubbling dry HCl gas into anhydrous methanol (Stoffel et al., 1959). Methylation using 5% (w/v) methanolic HCl is complete in 1 h at 80°C for all lipids including *N*-acyl lipids (Stoffel et al., 1959; Christie, 2003; Kramer et al., 1997 & 2001). Methanolic HCl can also be produced by adding acetyl chloride into dry methanol (Lepage and Roy, 1986). However, methylation reactions may not be complete if the reagent is not dry as evidenced by using a 0.5M HCl solution (Murrietta et al., 2003). Other acid-catalyzed methylating reagents include H_2SO_4 in dry methanol and BF_3/methanol (Christie, 2003). The latter should only be used when it is freshly prepared because it has a limited shelf-life, and may result in the production of artifacts and loss of PUFAs (Christie, 2003; Aldai et al., 2005). BF_3 also results in partial or complete destruction of the following groups: epoxy, hydroperoxy, cyclopropenyl, cyclopropyl, and possibly hydroxyl (AOCS Ce 2-66 and ISO 5509). The disadvantage of all acid-catalyzed methylation procedures (including isopropyl and butyl esters) is that they isomerize *c/t* and *c,c* to *t,t*-CLA (Kramer et al., 1997). In addition, hydroxy fatty acids, known to be present in milk fats, are converted to methoxy artifacts (Kramer et al., 1997; Yurawecz et al., 1994). Lowering the temperature to 60°C (Park et al., 2001), or room temperature (Werner et al., 1992), decreased CLA isomerization somewhat (Park et al., 2001), but under these milder conditions methylation may not be complete (Kramer et al., 1997).

Base catalysts are generally selective and do not convert all acyl moieties to FAME, such as FFA, *N*-acyl lipids and alk-1-enyl ethers (Kramer et al., 1997; Cruz-Hernandez et al., 2004 & 2006); see Table 2. This obviously has quantitative implications. Methylation using sodium methoxide (e.g., 0.5N methanolic base #33080, Supelco Inc., Bellefonte, PA) is preferred for the determination of ruminant derived fats that contain CLA. The conversion is rapid (15 min at 50°C) and does not cause isomerization of CLA (Kramer et al., 1997). A method was reported recently for the preparation of esters with longer chain alcohols (ethanol, n-propanol, n-butanol, or 2-methoxyethanol) using potassium tert-butoxide as proton exchange reagent (Destaillats and Angers, 2002a). For milk fat, a water-free system has been used extensively for the preparation of FAME to prevent the loss of short-chain FAME (Christie, 1982; Chouinard et al., 1999; Cruz-Hernandez et al., 2004). Another base-catalyzed procedure was proposed in the ISO 5509 method which involved the use of trimethylsulfonium hydroxide. This reagent was said to be applicable for milk fat and blends containing milk fat. However, the ISO 5509 method does not recommended this reagent for use with cyanopropyl siloxane columns. This appears somewhat contradictory since cyanopropyl siloxane columns are recommended for *trans* analysis in the ISO 15304 method. El-Hamdy and Christie (1993) evaluated this catalyst and showed that fatty acids form a

Table 2. The Suitability of Catalysts for the Methylation of Specific Lipid Classes, and for Treatment of CLA Containing Fatty Acids [Reproduced with permission by AOCS Press; Cruz-Hernandez et al., 2006]

Type of Lipids	Structures	Catalysts[a]				
		HCl	BF$_3$	TMG	NaOCH$_3$	DAM
Esters	RCH$_2$-CO-OR^1	Y	Y	Y	Y	No
Free Fatty Acids	RCH$_2$-COOH	Y	Y	Y	No	Y
Amides	RCH$_2$-CO-NHR^1	Y,L	Y,L	No	No	No
Alk-1-enyl ethers	RCH$_2$-O-CH=CHR^1	Y	Y	?	No	No
Phospholipid esters	R(PO$_4$X)(OOCR^1)	Y	Y	No	Y	No
Cholesterol esters	RCOO-cholesteryl	Y,L	Y,L	?	No	No
Ethers	RCH$_2$-O-CH$_2$R^1	No	No	No	No	No
Conjugated bonds[b]	RCH=CH-CH=CHR^1	Isom	Isom	Stable	Stable	Stable

[a] TMG, tetramethylguanidine; DAM, diazomethane (or TMS-DAM, trimethylsilyl diazomethane); L, longer reaction times and generally higher temperatures are required; No, catalyst not suitable; ?, reaction unknown.
[b] Fatty acids containing conjugated double bonds may be present as an ester, amide or ether; see suitability of respective lipid type above. The conjugated double bond system in the alkyl chain of CLA may be isomerized (Isom) or is stable (Stable) using the methylation conditions.

complex with trimethylsulfonium hydroxide that pyrolyzes in the heated injection port of the GC to give dimethylsulphide, methanol and FAME as end products. The FAME produced in the injection port are subsequently separated in the GC column. Furthermore, the reaction was shown to be incomplete for cholesteryl esters (El-Hamdy and Christie, 1993) and FFA (ISO 5509).

It is apparent from the chemical composition of some foods that there is no single methylation procedure that adequately addresses each of these different lipids; base-catalyzed methylation does not convert FFA, N-acyl lipids and plasmalogenic lipids, while acid-catalyzed methylation isomerizes conjugated fatty acids. Three approaches have been considered: *Option A.* Perform the two methylation procedures in sequences, i.e., first add 0.3 mL NaOCH$_3$/methanol and react the sample at 50°C for 15 min, followed by adding excess 5% HCl/methanol (1 mL) and heat at 80°C for 30 min (Kramer et al., 2007). The AOCS methods Ce 2-66 and Ce 1k-07 recommend NaOCH$_3$/methanol followed by BF$_3$. This combination of derivatization procedures was recommended to address the problem of different lipid classes, but quantitative analysis of all lipid classes has not been critically evaluated. *Option B.* Hydrolyze the lipids to FFA using NaOH/ethanol (50°C for 1 hr, or over-night at RT) followed by methylation of the FFA using trimethylsilyl-diazomethane (TMS-DAM). Complete hydrolysis of all lipids with alkali is doubtful (Cruz-Hernandez et al., 2004), and subsequent methylation with diazomethane reagents has been reported by some to result in extra peaks (Park et al., 2001). *Option C.* Perform two separate methylation procedures, i.e., react the sample with NaOCH$_3$/methanol at 50°C for 15 min, and a duplicate sample with HCl/methanol at 80°C for 1 hour. The two results are then merged using either native 16:0 or 18:0 in the sample as internal reference, or simply use the acid-catalyzed results for quantitative purposes and correct only the CLA region by substituting the correct CLA isomer distribution obtained from the base-catalyzed results (Cruz-Hernandez et al. 2006). In our experience, option C has proven to be most versatile and reliable, and it provides the added benefit that the analyst actually performs a duplicate analysis.

The question of whether a duplicate methylation procedure (option C) is required depends on the chemical composition of the test samples in question. In general, the duplicate methylation procedure is not necessary for partially hydrogenated fats and oils, fully refined vegetable oils, and plant fats since they do not contain CLA and lipids stable to base-catalyzed methylation. The small amounts of CLA isomers present in PHVO as the result of hydrogenation (Azizian and Kramer, 2005) and those present in refined vegetable oils as the result of deodorization (Destaillats and Angers, 2002b; Juanéda et al., 2003) are a random distribution of CLA isomers, and any further randomization of the CLA isomer as a result of acid-catalyzed methylation (Kramer et al., 1997) would not seriously affect the CLA isomer distribution. On the other hand, base methylation may be adequate for the analysis of milk and beef fat, since they contains CLA but no plasmalogenic lipids, and the small amount of sphingomyelin (about 0.3% of total lipids) in these fats could be ignored. However, meat and blood lipids definitely require the duplicate methylation analyses, since more than 10% of the total long-chain alkyl groups are non-acyl and would therefore be ignored (Horrocks, 1972). Several comprehensive analyses of meats have recently been reported that include detailed analysis of the CLA and most *trans*-18:1 fatty acid isomers, but not those of plasmalogenic lipids (Fritsche et al., 2001; Nuernberg et al., 2002; Bessa et al., 2005; Hristov et al. 2005; Leheska et al., 2008).

The application of two separate methylation procedures has made it possible to obtain the complete analysis of meat lipids that includes CLA, *trans* fatty acid, and the plasmalogenic lipids (Santercole et al., 2007; Cruz Hernandez et al. 2006; Kraft et al., 2008; Aldai et al. 2009). Briefly, base-catalyzed methylation of meat lipids produced only FAME (Fig. 5A), while acid-catalyzed methylation resulted in a mixture of FAME and DMA that are co-analyzed by GC (Fig. 5B). The chemical structures of alk-1-enyl ethers and their reaction with acid or base catalysts is shown in Fig. 6. The DMA are thermally unstable and will undergo loss of one methanol group in the injection port of the GC to form alk-1-enyl methyl ether (AME) (Fig. 7) (Mahadevan et al., 1967); the resultant AME occur in both *cis* and *trans* configuration (Stein and Slawson, 1966). For confirmation of the different FAME and DMA peaks, the mixture can be separated by TLC using 1,2-dichloroethane (Winterfeld and Debuch, 1966). GC analysis of the TLC fractions permits identification of the peaks associated with the FAME (Fig. 5C), and those associated with the DMA and AME (Fig. 5D). The elution order of a specific alkyl chain length on the 100 m CP Sil 88 column was AME before DMA followed by FAME. The DMA after TLC separation can be converted to their stable cyclic acetals (Fig. 7) (Venkata Rao et al., 1967) that can be analyzed by GC (Fig.5E), which avoids the thermal degradation of the DMA in the GC injector.

One issue that has never been addressed is how to deal with non-acyl moieties such as plasmalogenic lipids, ubiquitously found in all meats (Horrocks 1972). Acid-, but not base-, catalyzed methylation of plasmalogenic lipids produces DMA from the long-chain alk-1-enyl moieties, and these DMA are co-analyzed with the FAME by GC (Horrocks 1972; Cruz Hernandez et al. 2006; Santercole et al., 2007; Kraft et al. 2008). Consumption of plasmalogenic lipids easily convert alk-1-enyl moieties to the corresponding fatty acids in biological systems and they become part of the fatty acid pool. The question remains whether to include these alkenyl moieties into the total fatty acid content of a food product. The energy derived from these long-chain alkenyl moieties would certainly be indistinguishable from that produced from a fatty acid having the same chain length.

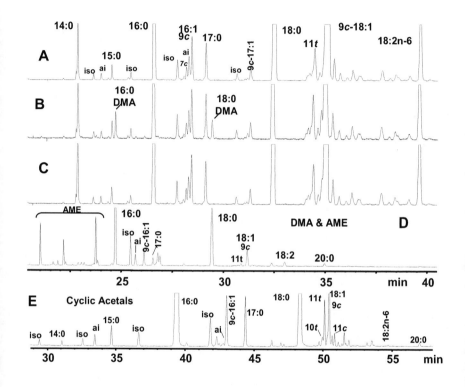

Fig. 5. Partial GC chromatograms of total sheep lipids methylated using a base- (A) or an acid-catalyzed procedure (B). The acid-catalyzed methylation mixture (B) was resolved by TLC (developing solvent, 1,2-dichloroethane) into the FAME derived from O-acyl lipids (C) and dimethylacetals (DMAs) from the plasmalogenic lipids (D). The DMA mixture was converted to the more stable cyclic acetals (E) used to determine the alk-1-enyl composition. All GC separations were performed using the same GC column and temperature program. The temperature of the injection port (250°C) caused partial loss of one methanol from the DMAs to form a mixture of alk-1-enyl methyl ethers (AMEs) and DMA. [Reproduced by permission of Lipids and authors; Santercole et al., 2006].

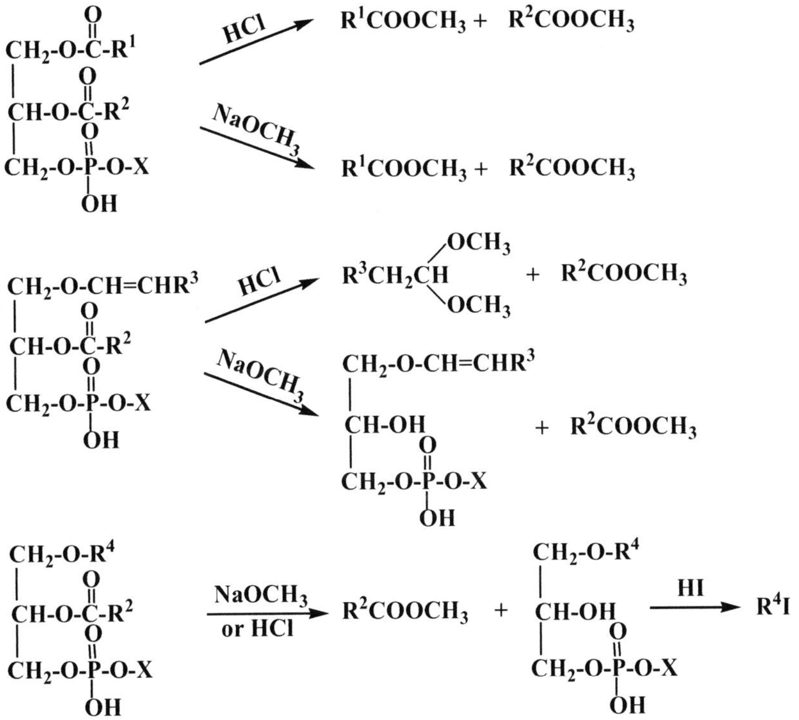

Fig. 6. Methylation of diacyl, alk-1=-enyl acyl and alky acyl phospholipids using an acid (HCl) or base (NaOCH3) catalyst. X = ethanolamine (PE) or choline (PC); R1 to R4 are different alkyl chains. [Reproduced by permission from AOCS Press and authors; Cruz-Hernandez et al., 2006].

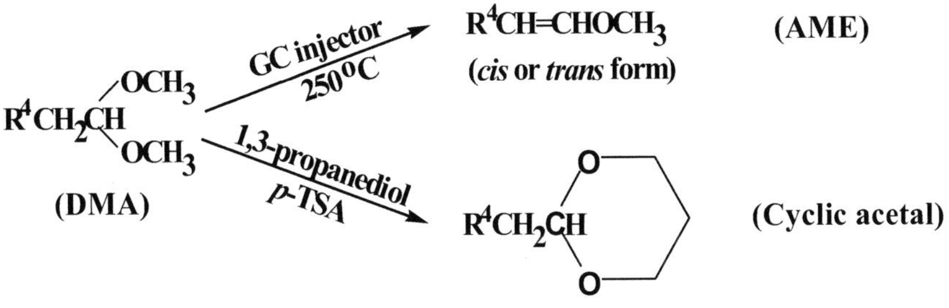

Fig. 7. There is partial loss of methanol from the dimethylacetals (DMA) under the heat conditions of the GC injection port resulting in the corresponding alk-1-enyl methyl ethers (AMEs). The latter occur in the *cis* and *trans* configuration. DMA can be converted to stable cyclic acetals (dioxolane derivatives) by reaction with 1,3-propanediol in the presence of p-toluenesulfonic acid (p-TSA). The cyclic acetals are stable under the GC injection conditions. [Reproduced by permission from AOCS Press and authors; Cruz-Hernandez et al., 2006].

Selection of GC column and column length

Currently, there is general agreement that the very long (100 m or longer) fused silica capillary columns coated with highly polar (100% cyanopropyl polysiloxane) stationary phases are mandatory for the *trans* fatty acid determination of fats and oils from all matrices (Molekentin and Precht, 1995; Precht and Molkentin, 1996, 1997, 1999, 2000c & 2003; Wolff et al., 1998; Ratnayake, 1998; 2001, 2004 & 2006; Kramer et al., 2001, 2002 & 2008; Wolff and Precht, 2002; Cruz-Hernandez et al. 2004 & 2006; Shingfield et al., 2006; Destaillats et al. 2007; Shirasawa et al., 2007; Rozema et al., 2008). The 50 or 60 m columns with highly polar stationary phases are not included in this discussion because these columns provide insufficient separation of many *trans* fatty acid isomers (Molkentin and Precht, 1995; Precht and Molkentin, 1997; Wolff and Precht, 2002). These shorter columns were used in previous official methods (AOCS Ce 1f-96; AOAC 996.06 - first action 1996; ISO 15304) for the analysis of fully refined vegetable oils (Fig. 8) and PHVO (Fig. 9). It was estimated that as much as 30% of the total *trans* fatty acid content was underestimation with these shorter capillary columns (Duchateau et al., 1996; Wolff and Precht, 2002; Golay et al., 2006).

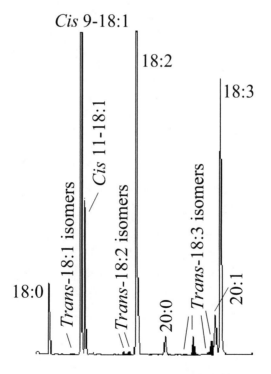

Fig. 8. Partial GC FAME profile (18:0 to 18:3) reported in AOCS method Ce 1f-96 for partially hydrogenated soybean oil using a 50-m CP Sil 88 column. [Reproduced by permission from AOCS Press].

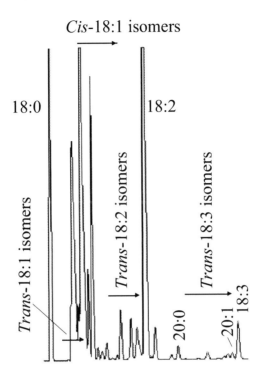

Fig. 9. Partial GC FAME profile (18:0 to 18:3) reported in AOCS method Ce 1f-96 for refined rapeseed oil using a 50-m CP Sil 88 column. [Reproduced by permission from AOCS Press].

The 100% cyanopropyl polysiloxane stationary phases are marketed as CP-Sil 88 (Varian Inc., made by Chrompack Inc., Middelburg, The Netherlands), SP-2560 (Supelco Inc.), and Rtx-2560 (Restek). Since these columns are not bonded, their maximum temperature is about 225°C, and rinsing with organic solvents should be avoided. Typical separations in the 18:1 region are shown for a PHVO (Fig. 10) and a milk fat (Fig. 11). The 100 m CP Sil 88 or SP2560 also improved the separation in the 16:1 (Precht and Molkentin, 2000a; Kramer et al., 2008), 18:2 (Precht and Molkentin, 1997, 2003; Kramer et al. 2008), 20:1-18:3 (Wolf 1994; Precht and Molkentin, 1999; Kramer et al., 2008) and CLA regions (Roach et al., 2000 and 2002; Cruz-Hernandez et al. 2004 & 2006; Shingfield et al., 2006).

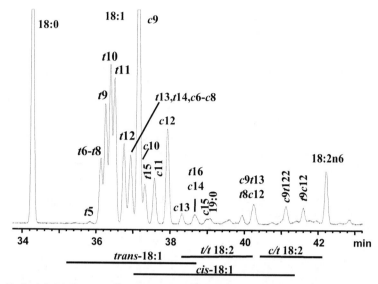

Fig. 10. Partial GC FAME profile (18:0 to 18:2) for a partial hydrogenated vegetable oil using a 100-m CP Sil 88 column and a temperature program from 45 to 215°C (Kramer et al., 2001 and 2002). The partial overlap of *trans*-18:1, *cis*-18:1 and *cis/trans* 18:2 isomers is indicated. [Reproduced by permission from AOCS Press and authors; Mossoba et al., 2003 and 2005].

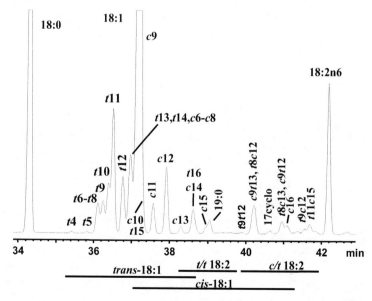

Fig. 11. Partial GC FAME profile (18:0 to 18:2) for total milkfat using a 100-m CP Sil 88 column and a temperature program from 45 to 215°C (Kramer et al., 2001 and 2002). The partial overlap of *trans*-18:1, *cis*-18:1 and *cis/trans* 18:2 isomers is indicated. [Reproduced by permission from AOCS Press and authors; Mossoba et al., 2003 and 2005].

There are several suppliers of very long capillary columns that have a slightly lower polarity, such as HP 88 (Agilent Technologies), BPX-70 (SGE Inc.) and TC-70 (GL Sciences Inc.; Shirasawa et al., 2007). The HP 88 column is a cyanopropyl polysiloxane stationary phase (88%) stabilized by incorporating arylene bridges into the polymer backbone (Vickers et al., 2004). This bonding allows for greater stability of the column. The BPX-70 column is a bonded cyanopropyl polysilphenylene siloxane polymer phase equivalent to 70% cyanopropyl polysiloxane column, and is available in 60 m and 120 m lengths. The TC-70 is a 70% cyanopropyl (equivalent)-silphenylene-siloxane column available in 60 m length. Based on the few available published separations of the *trans* isomer region, these columns appears to be less efficient by comparison to the 100% cyanopropyl polysiloxane stationary phases (HP 88: Vickers et al., 2004; BPX-70: Destaillats and Angers, 2002b, Destaillats et al., 2005; ISO 15304; TC-70: Shirasawa et al., 2007). A thorough evaluation of the separations of each of these columns, including the 120 m BPX-70 column, is required to establish the elution pattern of most of the common FAME including the 18:1, 18:2 and 18:3 isomers, and the identity of possible interfering FAME in the CLA region. Only partial separations of the CLA (Destaillats and Angers, 2002b & 2003; Destaillats et al., 2005) and *trans*-18:1 isomers (Juenéda, 2002; Destaillats et al., 2005) have been published using the 120 m BPX-70 columns.

In the last few years, a 200 m chemically bonded capillary column has also become available (CP SELECT for FAMES, Varian Inc.; Peere et al., 2004). Even though this column is of intermediate polarity, the increased length provides separations of the 18:1 isomer region similar to those observed using the 100% cyanopropyl polysiloxane stationary phases (Fig. 12; courtesy G. Jahreis and P. Möckel). However, using such a column requires higher head pressures in the GC, which makes the junction fitting of two 100m columns and the column at the injector end more prone to leakage (Möckel P and Jahreis G, private communication). Details on the separations and identifications of FAME in all the different regions remain to be established.

Limitation of 100 m GC columns

The attempt to select an optimum GC program for use as the official method to determine all the *trans* fatty acid isomers in food products has its limitation. Despite the marked improvement in the resolution using these 100 m, highly polar columns, even the separation of the geometric 18:1 isomers remains a challenge that requires the sorting out of overlapping peaks for *cis* and *trans* isomers. The extent of the overlap of the 18:1 geometric isomers is evident after a prior fractionation of the geometric isomers using silver-ion chromatographic techniques followed by GC separations (Fig. 13). The *trans*-18:1 isomers eluted prior to the corresponding geometric *cis* isomer, with peaks for *c*6-*c*8- to *c*14-18:1 isomers overlapping with those for *t*13/*t*14- to *t*16-18:1 isomers. Generally, all *trans* products (PHVO, deodorized vegetable oils, oils used in high temperature frying, and ruminant fats) contained the same geometric 18:1 isomers, but they differed in their relative

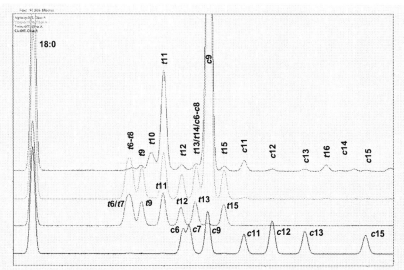

Fig. 12. Separation of the 18:1 region using a 200 m CP Selct column operated using an isothermal conditions at 180°C. The chromatograms from top to bottom are: total milk fat FAME; *trans*-18:1 standard mixture plus methyl oleate (c9-18:1); *trans*-18:1 standard mixture; and *cis*-18:1 standard mixture. The chromatogram of total milk fat is labeled separately because some isomers may overlap or contain more isomers than present in the standard mixtures. The isomers present in the *trans* and *cis* mixtures are identified in the lower two chromatograms. [The authors thank Dr. G. Jahreis and Mr. P. Möckel, University of Jena, Germany for making available these GC separations].

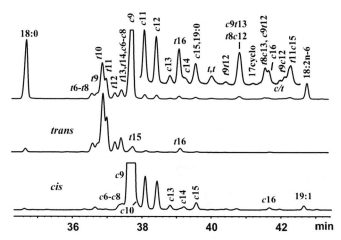

Fig. 13. Partial GC FAME profile (18:0 to 18:2) for total milkfat, and the *trans* and *cis* fractions isolated by silver ion-SPE. A 100-m CP Sil 88 column was used, and a temperature program from 45 to 215°C (Kramer et al., 2001 and 2002). [Reproduced by permission of Lipids and authors; Kramer et al. 2008].

abundance (Molkentin and Precht, 1995; Wolff et al., 2000; Precht and Molkentin, 1997; Cruz-Hernandez et al., 2004 & 2006). The quality of the separation and resolution of the individual peaks for 18:1 isomers will depend on the relative abundance of the peaks for adjacent isomers (Cruz-Hernandez et al., 2004; Kramer et al. 2008). For example, in PHVO there is generally a near baseline resolution of peaks for *t*13-/*t*14-18:1 and *c*9-18:1, and the peak for *t*15-18:1 eluted shortly after that of *c*9-18:1 (Fig. 10), but such a separation is not generally possible with milk fat using the same GC conditions, because of larger differences in concentration of adjacent isomers (Fig. 11). To better resolve and identify these geometric isomers, advantage was taken of the fact that the polarity of cyanopropyl phases was shown to be temperature dependent (Castello et al. 1997; Mjøs 2003). A change in operating temperature of the GC column results in differences in the elution times of peaks for fatty acids of the same type, but more importantly, there were differences in the relative elution of peaks for different fatty acid types (Precht & Molkentin 1999 & 2003; Kramer et al, 2008). For instance, the relative separation of the *cis*- and *trans*-18:1 isomers was significantly affected by temperature that resulted in different patterns for the 18:1 isomers (Fig. 14; Kramer et al., 2008).

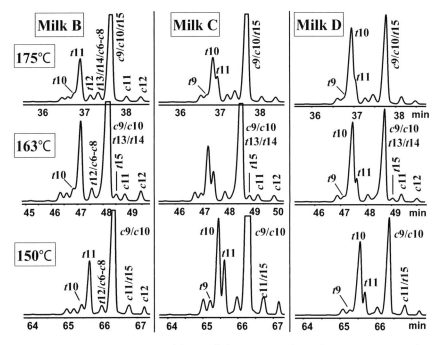

Fig. 14. Partial GC chromatogram of three milk fats (B-D) are shown from *t*4- to *c*12-18:1 that contained a high total *trans*-18:1 content. Separations were performed using the 175°C (top row), 163°C (middle row) and 150°C (bottom row) GC programs. Selected *trans*- and *cis*-18:1 isomers are identified to indicate the changes in the elution order of these 18:1 isomers with temperature. [Reproduced by permission of Lipids and authors; Kramer et al. 2008].

Three different milk fats were separated at three different GC temperature programs that plateaued at 175°C, 163°C and at 150°C (Fig. 15–17) (Kramer et al. 2008). It appears that the 163°C program might be ideal to use as a single GC temperature program. However, this would only be based on the separation of *t*15-18:1. A larger temperature shift is required to provide sufficient differences in elutions between different lipid types to be useful for identification. The drop in temperature from 175°C to 150°C resulted in an improved separation of the peaks for *t*6-*t*8- to *t*11-18:1 isomers (Fig. 14), and moved the peak for *t*15-18:1 that coeluted with that of *c*9-18:1 at 175°C (Fig. 15) to the retention time of the peak for *c*11-18:1 at 150°C (Fig. 17); while at 163°C this peak for the *t*15-18:1 isomer was complete resolved (Fig. 16). Separation of all the *trans*-18:1 isomers was only possible at 120°C and included the difficult resolution of the two normally co-eluting peaks for the *t*13-18:1 and *t*14-18:1 isomers, as demonstrated by using the isolated *trans* fraction by silver-ion SPE (Fig. 18).

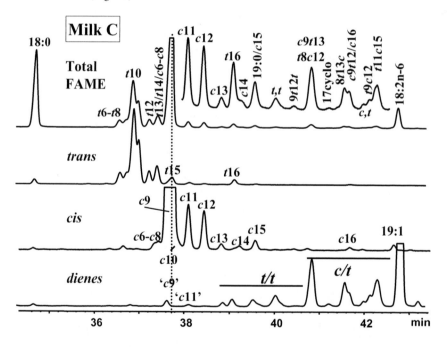

Fig. 15. Partial GC chromatogram of milk fat C separated using GC program 175°C and showing the 18:0 to 18:2n-6 region. An enlargement of the *c*11-18:1 to *t*11*c*15-18:2 is inserted for clarity. The *trans*, *cis* and diene fractions isolated from total milk fat FAMEs using Ag⁺-SPE columns and are compared to the total milk fat FAME. The structure of all the *t/t*-18:2 isomers, except t9t12-18:2 is unknown. The assignments of most c/t-18:2 isomers and methyl 11-cyclohexylundecanoate (17cyclo) is tentative, and based on comparison with some standards (*c*9t12-18:2 and *t*9c12-18:2) and on previously published reports. [Reproduced by permission of Lipids and authors; Kramer et al. 2008].

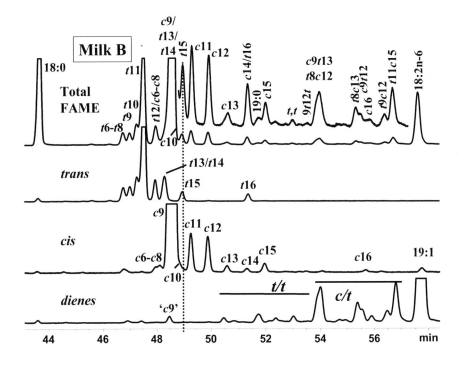

Fig. 16. Partial GC chromatogram of milk fat B separated using GC program 163°C and showing the 18:0 to 18:2n-6 region. An enlargement of the *c*11-18:1 to *t*11*c*15-18:2 is inserted for clarity. For remaining description see caption Fig. 15. [Reproduced by permission of Lipids and authors; Kramer et al. 2008].

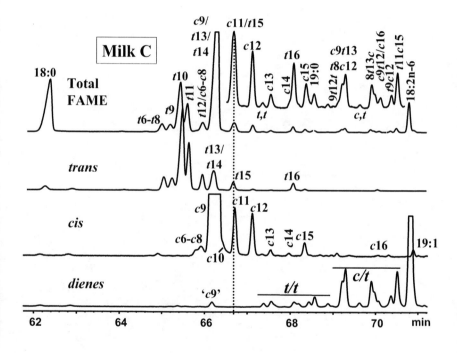

Fig. 17. Partial GC chromatogram of milk fat C separated using GC program 150°C and showing the 18:0 to 18:2n-6 region. An enlargement of the c11-18:1 to t11c15-18:2 is inserted for clarity. For remaining description see caption Fig. 15. [Reproduced by permission of Lipids and authors; Kramer et al. 2008].

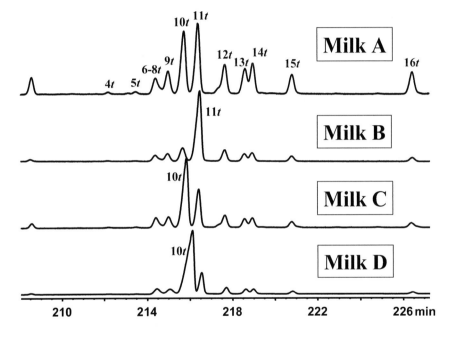

Fig. 18. Partial GC chromatogram of the 18:1 region of the *trans* fraction isolated from all four milk fats selected (A-D) using Ag⁺-SPE chromatography and analyzed by GC operated isothermally at 120°C. [Reproduced by permission of Lipids and authors; Kramer et al. 2008].

Changes in operating temperature of the GC column were also used to resolve the FAME peaks in a number of other regions of the chromatogram. For examples, the coeluting peaks for *cis*- and *trans*-16:1 isomers together with those for the branch-chain 17:0 isomers in milk fat were resolved; this was determined by comparing the separations using temperature programs at 175°C and 150°C (Kramer et al., 2008). Precht and Molkentin (2003) resolved several *c/t*-, *t,t*- and *c,c*-18:2 isomers that coeluted with *cis*-18:1 isomers, 19:0, 19:1, and a cyclic fatty acid (11-cyclohexylundecanoic acid) in milk fat by conducting isothermal operations every 10°C between 130°C to 190°C "(it is nor clear whether more than one column from each make were used). Furthermore, 20:1 was resolved from the geometric isomers of 18:3n-3 in fats and oils using isothermal temperatures from 155°C to 180°C (Wolff, 1994). Different temperature programs were used to resolve the positional and geometric isomers of 20:1 from the geometric isomers of 18:3n-3 (Kramer et al., 2008). However, the relative separation of the different CLA isomers was not altered by changes in the temperature settings (Kramer et al., 2008), and even at isothermal 120°C the separation of the CLA isomers did not improve (Kramer et al., 2001). However, the elution of 21:0 relative to the CLA isomers changed with changes in column temperature, eluting with *t*9*c*11-18:2 at 175°C and shortly after *t*10*c*12-18:2 at 150°C (Kramer et al., 2008).

Generally, different GC columns of different polarities have been shown to have a high diagnostic capability that is valuable for the identification of unknowns. It would be interesting to compare several different columns of slightly different polarities, such as the 100 m HP 88, 120 m BPX-70, or the 200 m CP SELECT for FAMES columns, to complement the resolution of all the geometric and positional isomers obtained by using the CP Sil 88 and SP 2560 columns..

Differences between 100 m cyanopropyl polysiloxane GC columns

Theoretically, columns with the same stationary phase and the same length should give similar separations. However, there is some variability due to differences in the construction of the columns. For instance, some suppliers produce continuous 100 m capillary columns (i.e., SP-2560; Rtx 2560; HP 88), while others join two 50 m capillary columns (i.e., CP Sil 88), or two 100 m capillary columns (200 m CP SELECT). Ratnayake's group extensively investigated the separation of the FAME from a PHVO source using a 100 m SP2560 and a CP Sil 88 column both operated at isothermal temperature from 170°C to 190°C (it is not clear whether more than one column from each make were evaluated). They observed several minor differences between these two columns, and concluded that isothermal separations at 180°C provided the best overall separation of most *trans* fatty acids (Ratnayake et al., 2002 & 2006). Furthermore, this is not clear from these reports whether more than one GC column was tested from each supplier. This is important because batch differences have been observed between columns that cannot be corrected by small incremental adjustments of 1°C as proposed in AOCS 1h-05; see discussion below. Previous analyses had concluded that the separation obtained with these columns was optimum at 175°C isothermal condition (Molkentin and Precht, 1995; Precht and Molkentin, 1997). Official methods encourage the analysts to obtain similar separations to those published, and suggest that minor adjustments should be made in the column oven temperature by incremental steps of 1°C (AOCS Official Method Ce 1h-05; AOCS Recommended Practice Ce 1j-07; ISO 15304).

The assumption is generally made that the chemical composition of the 100% cyanopropyl polysiloxane stationary phase is identical in all columns, which may not be the case. We have previously noted that the elution pattern of the FAME is slightly different between identical makes of columns purchased from the same supplier and operated using the same temperature program. This was evident in the elution pattern of 21:0 relative to the CLA isomers from before *c*11*t*13-18:2 to slightly after *t*10*c*12-18:2, as seen in Fig. 19 (Cruz-Hernandez et al., 2004 & 2006). Both differences between columns and age of the column influenced the relative elution of these FAME. We have also observed differences in the relative

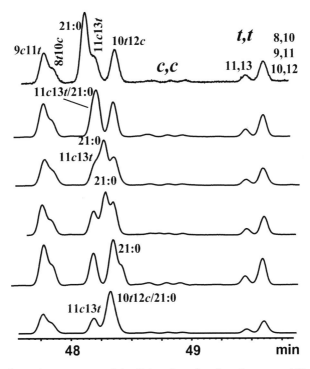

Fig. 19. Partial GC chromatograms of the CLA region taken from 3 separate 100 m CP Sil 88 columns and at different times of the life of the GC column. The standard CLA mixture from Nu-Chek Prep contained 4 positional isomers and was spiked with the methyl ester of 21:0. The GC separations show the differences one observes in the relative elution order of the CLAs and 21:0. [Reproduced by permission from AOCS Press and authors; Cruz-Hernandez et al., 2006].

elution pattern of other FAME using the same column make and supplier over a 3 year period (Kramer unpublished data). These differences cannot be eliminated by incremental changes in the column oven temperature as suggested in the official methods Ce 1h-05, Ce 1j-07, and ISO 15304. Therefore, the analyst is cautioned to thoroughly evaluate each new GC column for its separation capability and elution pattern before use.

Effect of carrier gas

Official methods generally recommend the use of either hydrogen or helium as carrier gas. Evaluations have generally shown that hydrogen provides better resolution of all the *trans* fatty acids (Kramer et al., 2002; Ratnayake et al., 2006; Rezema et al. 2008) partly due to its higher thermal conductivity (Grop and Biedermann, 2003). In the author's opinion (JK), the advice to avoid the use of hydrogen as carrier gas because of possible explosion is unfounded considering that hydrogen is also used as one of the gasses in FID detection, and current GC instruments are equipped with hydrogen sensors that turns off all hydrogen flows once a leak is detected. However, vigilance is always necessary to insure that the gas lines to the GC do not leak.

Effect of sample load

Overloading a capillary column will markedly affect the resolution of closely eluting adjacent isomers. This is a particular concern for the *trans* fatty acids since they elute in close proximity. Some separations can be improved by reducing the sample load such as the resolution of the *t*13-/*t*14-18:1 peak before *c*9-18:1 (Fig. 10 and 11). However, other separations cannot be attained even at reduced sample load because of the relative concentration of the adjacent isomers. For example, *t*15-18:1 can be resolved from *c*9-18:1 in most PHVO (Fig. 10), but not in milk fats, unless the temperature conditions of the column is reduced from 175°C to 163°C (Fig. 16; Kramer et al., 2008). The same applies to the separation of the *trans*-18:1 isomer from *t*6-*t*8- to *t*11-18:1. In PHVO these four peaks are generally in similar abundance, which permits their estimation by dropping perpendiculars (Fig. 10). However, in milk fats the distribution of isomers is uneven (Fig. 11 and 13). A greater separation of these isomers was achieved for more definitive identification and quantitation by analyzing the samples at different column temperatures (Fig. 14; Kramer et al., 2008). The separation of *t*16-18:1 and *c*14-18:1 is another example of an isomer pair that often overlaps. On the other hand, all attempts have failed to resolve some isomer pairs such as *t*7,*c*9- from *c*9,*t*11-18:2.

Complementary silver-ion separations of geometric isomers

To overcome the problem of overlapping geometric isomers, prior separations of the *cis* and *trans* isomers were conducted using silver ion-TLC (Henninger and Ulberth, 1994; Chen et al. 1995; Wolff, 1995; Precht and Molkentin, 1996, 1999 & 2000c; Chardigny et al., 1996; Kramer et al., 2001; Cruz-Hernadez et al., 2004 & 2006), silver ion-HPLC (reviewed by Ratnayake, 2004), or silver ion-SPE

(Kramer et al. 2008). GC analysis of the isolated *trans* and *cis* fractions confirmed the extent of overlap (Fig. 13, 15–17). Conducting the GC analysis of the isolated *trans* fraction at an isothermal temperature of 120°C resolved all the individual 18:1 isomers including the *t*13- and *t*14-18:1 pair (Fig. 18) that could not be revolved at the higher temperatures (Precht and Molkentin, 1997, 1998, 1999, & 2000c; Mossoba et al., 1997; Kramer et al., 2001; Cruz-Hernandez et al., 2004 & 2006). With slight modification, the silver ion-SPE technique (Supelco Technical Report, 2006) proved highly successful for the isolation of the different fractions based on the number of double bonds and their geometric configuration (Kramer et al., 2008); see Figures 13 and 15–17.

Resolution of trans isomers other than 18:1

A similar overlap pattern of geometric isomers was evident in the 16:1 retention time region for ruminant fats that coeluted together with the C17 branch-

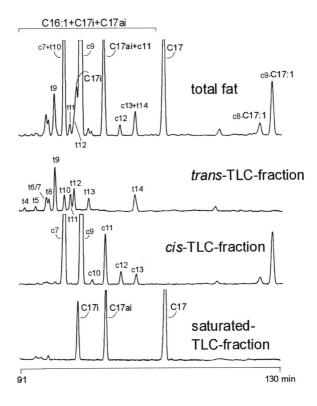

Fig. 20. Partial chromatograms of the 16:1 region comparing total milk fat FAME with fractions obtained following Ag⁺-TLC separation. A 100-m CP-Sil 88 column was used operated at 120°C isothermal conditions. [Reproduced with kind by permission from the European Journal of Lipid Science and Technology and the authors; Precht and Molkentin, 2000].

chain fatty acids; see Fig. 20 (Precht & Molkentin 2000a; Kramer et al. 2008). Fortunately, PHVO and fully refined vegetable oils contain only trace amounts of *trans*-16:1 formed during hydrogenation and deodorization, since plant oils contain only few unsaturated C16 fatty acids present in small amounts. This is in contrast to milk and ruminant fats that contain significant amounts of *trans*-16:1 isomers. The 20:1 region also consists of overlapping geometric isomers of 20:1 that coelute with the geometric isomers of linolenic acid (Wolff 1994; Precht & Molkentin 2000c; Kramer et al 2002 & 2008); see Figure 21 (Kramer et al., 2002). A decrease in the oven temperature from 180°C and at 155°C was shown to resolved the *cis*-20:1 from the *c/c/t*-18:3 isomers in PHVO and vegetable oils, thus permitting a better identification of these fatty acids (Wolff 1994). Similar improvements were observed by analyzing milk fats at 175°C and 150°C to resolve the *trans*- and *cis*-20:1 isomers; only trace amounts of *c/c/t*-18:3 isomers were found in milk fats (Kramer et al., 2008). Another region of equal complexity is the *cis*-18:1, *trans,trans*-18:2 and *cis/trans*-18:2 region that contain many isomers, some of which still remain unidentified (Precht & Molkentin, 1997 & 2003; Ratnayake, 1998 & 2002; Kramer et al. 2008; Ratnayake & Cruz-Hernandez, 2009); see also the chromatograms on the bottom of Figures 15 to 17.

Separation of CLA isomers using GC and the complementary Ag-HPLC techniques

On the 100 m cyanopropyl polysiloxane (100%) GC columns the CLA isomers elute between 18:3n-3 and 20:2n-6 (Fig. 21). The GC elution order of the CLA isomers is *c/t*- > *c,c*- > *t,t*-CLA as demonstrated using a commercial CLA mixture containing four positional isomers (Fig. 22, upper graph). Differences in the operating temperature of the GC columns does not affect the relative elution order of the CLA isomers (Kramer et al., 2008). Very few of the *c/t*- and *c,c*-CLA isomers overlap, an example being *t*11*c*13-CLA and *c*9*c*11-CLA. However, there are many CLA isomers that are not resolved, for example *t*7*c*9-CLA and *c*9*t*11-CLA and all the *t,t*-CLA isomers from *t*7*t*9-CLA to *t*10*t*12-CLA (Fig. 22). The CLA region is generally free of interfering fatty acids, except 21:0 and several *c/t*-20:2 isomers. The identities of CLA isomers and the other coeluting fatty acids were definitively confirmed using high resolution GC-mass spectrometry as shown in Fig. 23 (Roach et al. 2000 and 2002).

A complementary analysis was developed using silver-ion HPLC (Ag⁺-HPLC) with three ChromSpher 5 Lipids (250 × 4.6 mm stainless steel, 5μm particle size) columns in series that resolved most of these positional and geometric CLA isomers, which were detected by their UV absorption at 233 nm (Sehat et al. 1999; Kramer et al., 1999; Cruz-Hernandez et al. 2004). This method resolved the CLA isomers in the order: *t,t*- > *c/t*- > *c,c*-CLA; see typical separation of the commercial CLA mixture containing four positional isomers (Fig. 22, lower graph). The observed elution order within each group of geometric CLA isomers increases as the Δ positions decreases. For the same positional isomer with double bonds closer to

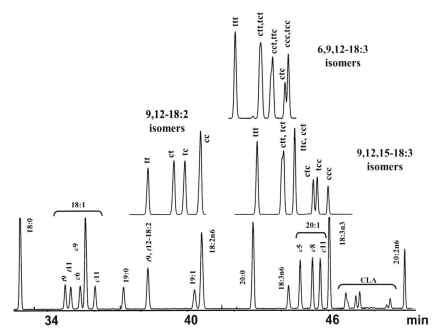

Fig. 21. Partial gas chromatogram of the 18:0 to 20:2n-6 FAME region using a 100-m CP Sil 88 capillary column, hydrogen as carrier, and a typical temperature program from 45 to 215°C. Lower panel: FAME standard #463 spiked with the CLA mixture #UC-59M, both from Nu-Chek-Prep (Elysian, MN); the isomeric mixture of linoleic (9,12-18:2), α-linolenic (9,12,15-18:3) and γ-linolenic acid (6,9,12-18:3) are placed above the lower chromatogram in the appropriate elution positions. [Reproduced by permission of Lipids and authors; Kramer et al. 2002].

the carboxyl group than 10,12-18:2, the *c,t* isomer elutes before the *t,c* isomer, and for those with double bonds further from the carboxyl group than 10,12-18:2, the elution order is reversed (Kramer et al., 1998). The elution order of the all the CLA isomers from 6,8- to 13,15-CLA was confirmed using synthetic CLA preparations (Delmonte et al., 2004 ; Destaillats and Angers, 2003).

A complete analysis of all the CLA isomers in ruminant fats requires a combination of GC and Ag⁺-HPLC results as shown in Fig. 24. The CLA peaks are quantitated by GC analyses of total FAME and the resolution obtained by Ag⁺-HPLC is applied to determine the relative CLA distribution in the GC results. There are four GC peaks in the CLA region that may cause problems of identification: the "*c*9*t*11-CLA" peak may contain *t*7*c*9- , *c*9*t*11- and *t*8*c*10-CLA (generally *t*8*c*10-CLA being a minor component), the "*t*10*c*12-CLA" isomer often includes 21:0, the "*c*9*c*11-CLA" peak may contain *t*11*c*13-18:2, and the "*t*9*t*11-CLA" peak is a mixture of several *t,t*-CLA isomers from *t*7*t*9- to *t*10*t*12-CLA. On the other hand, there are several CLA isomers that are resolved by GC, such as

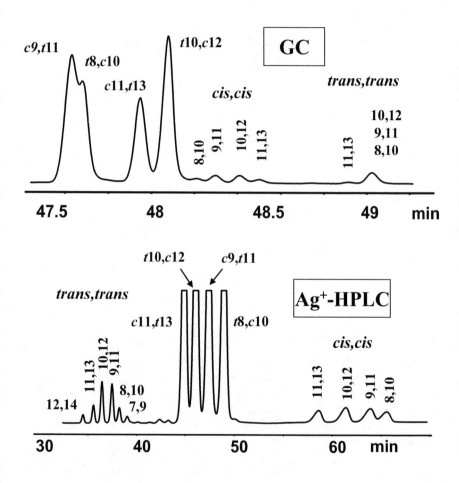

Fig. 22. Partial GC (upper) and Ag⁺-HPLC (lower) chromatograms of the CLA mixture #UC-59M from Nu-Chek-Prep (Elysian, MN). [Partial figures reproduced with kind permission from J AOAC International and authors; Cruz-Hernandez et al., 2004].

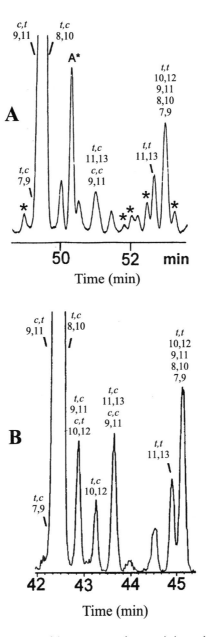

Fig. 23. Partial GC chromatogram of the CLA region from total cheese fat FAME analyzed using a 100 m CP Sil 88 capillary column (upper). The lower panel shows the same CLA region but analyzed by GC-MS using high-resolution select-ion recording at *m/z* 294.2559. A* is C21:0; * are non-CLA isomers. [Reproduced with kind permission from Lipids and authors; Roach et al., 2000].

CLA FAME region : GC and Ag⁺-HPLC

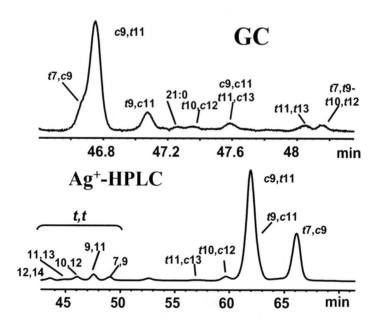

Fig. 24. Comparison of the CLA region of total milk fat FAME from cows fed a control diet using GC (A) and Ag⁺-HPLC (B). [Partial figure reproduced with kind permission from J AOAC International and authors; Cruz-Hernandez et al., 2004].

*t*9*c*11-, *c*10*c*12-, *c*11*c*13-, and *t*11*t*13-CLA. By combining these results it is possible to determine *t*7*c*9- CLA from Ag⁺-HPLC , *t*9*c*11-CLA by GC, and *c*9*t*11-CLA by difference (Fig. 24). The coeluting pair of 9*c*11*c*- and 11*t*13*c*-CLA as well as all the *t,t*-CLA isomers observed by GC are well resolved by Ag⁺-HPLC. Identification of 21:0 poses a challenge, since it generally occurs at a similar concentration as the minor CLA isomers, and elutes anywhere from before *c*11*t*13- to *c*10*c*12-CLA depending on the GC column and the temperature program used (Fig. 19). Therefore, it is essential to include both 21:0 FAME and the four positional CLA isomeric mixture into GC standards, such as #463 from Nu-Chek Prep (Kramer et al., 2002; Cruz-Hernandez et al., 2004 and 2006). Using two temperature programs also helps to identify 21:0, since it elutes later at the lower temperature relative to the CLA isomers (Kramer et al., 2008). The elution pattern of all the synthetic CLA isomers from 6,8- to 13,15-18:2 were resolved by Ag⁺-HPLC (Delmonte et al., 2005).

The Ag⁺-HPLC method has proven to be mandatory to adequately resolve all the positional and geometrical CLA isomers present in milk and meat fat of ruminants (Sehat et al. 1999; Cruz-Hernandez et al. 2004). An example of the complexity of the CLA region is shown in Fig. 25, which exhibits the GC and Ag⁺-HPLC separations of the CLA isomers of two milk fats from cows fed a CLA

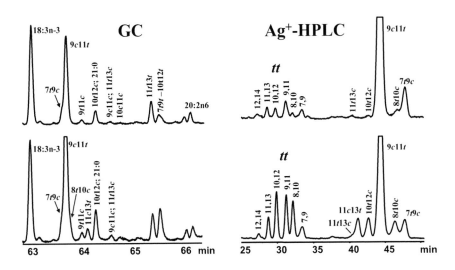

Fig. 25. Comparison of the CLA isomer regions separated by GC (left) and Ag⁺-HPLC (right) of commercial milk fat (upper) and milk fat from cows fed the Ca salt of CLA acids for 19 days (lower) (sample courtesy M. Sippel and Dr. J.P. Cant). These two complementary methods are recommended for the identification and quantitation of the CLA isomers in ruminant fats. [Reproduced by permission of AOCS Press and authors; Cruz-Hernandez et al., 2006].

mixture. CLA was added to the cows' diet to reduce milk fat; the latter was due to the presence of *t*10*c*12-18:2 (Bauman and Griinari, 2001). The CLA isomeric composition evident in this milk fat also points to the need to report rumenic acid separately, because of the amount and large distribution of non-rumenic acid *trans*-CLA isomers found in these mixtures. A similar array of CLA isomers were observed in retail beef stakes surveyed recently (Aldai et al., 2009). This new capability to separate individual *trans*-CLA isomers should allow the evaluation of the physiological effects of each one. Further research may not justify the exclusion of all of the *trans*-CLA isomers from *trans* labeling rules.

GC standards for FAME identification

GC standards are essential for the identification of fatty acids, for checking the reproducibility and efficiency of GC columns, and for quantitating the fat content in a given product. Many reference standard FAME mixture are currently available for identification of fatty acids, including product specific mixtures. The FAME mixture #463 from Nu-Chek Prep (Elysian, MN) is particularly useful since it consists of 53 FAMEs from C4 to C24, with up to six double bonds, and several *trans* geometric and positional fatty acids isomers. However, this mixture lacks CLA and several SFAs (21:0, 23:0 and 26:0) necessary for ruminant fat analysis. For this reason, we consistently add the SFAs 21:0, 23:0 and 26:0, and the CLA mixture #UC-59M from Nu-Chek Prep that consists of four positional CLA isomers (*t*8*c*10-, *c*9*t*11-, *t*10*c*12- and *c*11*t*13-18:2) and their corresponding *c*/*c* and *t*/*t* geometric CLA isomers (Kramer et al., 2001, 2002 and 2008; Cruz-Hernandez etal., 2004, and 2006). Several pure CLA standards (*c*9*t*11-, *t*9*c*11-, *c*11*t*13-, *t*10*c*12- and *t*9*t*11-18:2) are available from Matreya Inc. (Pleasant Gap, PA). Further confirmation of any FAME may be necessary by using argentation chromatographic techniques or GC/MS. The GC standard should also be used to check each column for reproducibility and separation characteristics.

A number of specific *trans* and *cis* octadecenoic acids and/or their methyl esters are available from Sigma Inc. (St. Louis, MO). A qualitative standard mixture of all the *trans* and *cis* isomers of 16:1, 18:1 and 20:1 can be prepared by isolating larger amounts of the *trans* and *cis* fractions from total milk fats by Ag^+-TLC or Ag^+-SPE. Some geometric isomers of 18:2 and 18:3 are commercially available. If not, they can be prepared by isomerization of methyl linoleate (*c*9*c*12-18:2; 18:2n-6), α-linolenate (*c*9*c*12*c*15-18:3; 18:3n-3) and γ-linoleate (*c*6*c*9*c*12-18:3; 18:3n-6) in dioxane at 100°C using *p*-toluenesulfinic acid as catalyst (Kramer et al., 2002).

Quantitation of fatty acid methyl esters by GC

Contrary to current opinion, GC is a secondary method and not a primary method of analysis, and is based on gravimetric preparation of a standard mixture of pure individual FAME. Currently, FAME mixtures, such as #714 from Nu Chek Prep, are being prepared for quantitative analyses that are accompanied with certificates of analysis. Ideally, such a mixture should be comprised of high purity FAME, which may not be possible. Analysts at Nu Chek Prep used a 20 m DB-23 capillary column to assess the purity of the FAME in the #714 quantitative mixture (private communication). The assessment of purity of each FAME should be conducted by GC (as opposed to other techniques), however with at least the power of resolution of GC columns used in the official methods. This raises a question about the accuracy of the certificate of analysis, since a 20 m DB-23 column will not resolve isomers that are separated using the 100 m capillary column specified in the official methods. Therefore, future preparations of quantitative FAME mixtures will need to be assessed for purity using the 100 m CP Sil 88 or SP 2560 capillary columns to provide a more accurate analysis of the standard for fatty acid determination. These mixtures are then used to determine the Empirical Correction Factors (ECFs), which is used for determining the response factor (area/amount) for each FAME in the mixture.

An additional concern in GC quantitation is the accuracy of introducing a FAME mixture from the injection port onto the column. On-column injection has been considered a more accurate means of quantitatively applying the sample onto the column, since this technique eliminates selective discrimination (Grob 2001; Grob and Biedermann, 2002). However, on-column injection is complicated and has therefore not been used extensively (Grop 2001). GC quantitation can be optimized by the selection of the appropriate mode of injection (splitt vs splitless), rate of injection, liners, sample volume injected, concentration of test sample, temperature program, and choice of carrier gas (hydrogen is preferred because of its high thermal conductivity; Grob and Biedermann, 2002). The development of product-specific methods should be considered. For instance, vegetable oil and their partially hydrogenated products have a narrow range of fatty acids, generally from C14 and C20. On the other hand, dairy and ruminant fats have a very broad range of fatty acids from C4 to C26. In addition, their methyl esters are prepared using $NaOCH_3$/methanol under anhydrous conditions with limited subsequent purification (Christie 1982; Chouinard et al., 1999; Cruz-Hernandez et al., 2004). Thus, there is a likelihood that traces of Na salts can be injected into the injection chamber and unto the column. Therefore, the total milk fat FAME mixture with a trace of contaminant will require a different set of injection conditions than the FAME mixture from vegetable oils; see principles expounded by Grob and Biedermann (2002). Generally, all methods call for a relative high concentration of the test mixtures (15–20 mg/mL) and the use of split injection. To date none of the official methods recommend different GC injection conditions. In the future, efforts should be made to develop product-specific methods that address concerns of selectivity and quantitation.

Experience

A successful GC determination of total *trans* fatty acids depends on the analyst's experience and skill in optimizing and evaluating the performance of the GC system, and subsequently identifying all the observed GC peaks in widely different and complex chromatographic profiles. Commercial FAME standards are often not available. The GC standard methods generally include the statement that peaks of unknown identity should not be included in the summation of total FAME unless the identities of such peaks are confirmed by using appropriate procedures involving either GC-MS, FTIR, silver-ion chromatography and/or classical chemical methods. In theory this is correct. However, in practice not all test samples can be rigorously evaluated, and sometimes these techniques are not readily available or accessible to everyone on a daily basis. Therefore, there might be a temptation to ignore unfamiliar peaks, which to an experienced analyst are not unknown peaks. This exclusion of unknown peaks compromises the accuracy of the analysis since the total of all known FAME peaks, and possibly the total *trans* fatty acid content may be different from the true values.

Official Methods

The most current official methods available for the determination of the *trans* fatty acids are summarized in Table 3. There are many common elements and procedures in these official methods, but also differences. The AOAC 996.06 is a more general method designed to determine the fatty acid composition of all foods including dairy and cheese, while the AOCS (Ce 1f-96, Ce 1h-05 and Ce 1j-07), ISO (15304), and the proposed JOCS (Kyo-4-1996) methods are recommended only for the determination of the *trans* fatty acids in partially hydrogenated fats and oils, and in fully refined and deodorized fats and oils. The AOAC method includes the preparation of the fat and oil samples and the subsequent methylation procedure, while the AOCS, ISO and the JOCS methods are essentially GC methods that require complimentary methods for the preparation and methylation of the samples, such as AOCS method Ce 2-66 (re-approved in 1997), Ce 1k-07 (approved 2007), ISO-661 and ISO-5509, and JOCS 2.4.1.2-1996. All these methods can also be used to obtain the full fatty acid composition in foods, including the total levels of saturated fatty acid (SFA), mono-unsaturated fatty acid (MUFA), and poly-unsaturated fatty acid (PUFA). Below, each of the steps in the various methods will be evaluated in turn.

Hydrolysis/saponification and methylation of plant derived fats and oils

The AOAC method 996.06 is the only method that calls for a prior acid hydrolysis of all non-ruminant products to their free fatty acids (FFA); see also House (1997). The conditions specify the addition of 8.3M aqueous HCl and heating at 70-80°C for 40 min. Acid treatment may lead to the partial or complete decomposition of functional groups such as, epoxy, hydroperoxy, cyclopropenyl, cyclopropyl and possibly hydroxyl and acetylenic fatty acids. For this reason a prior acid treatment is not included for plant derived fats in the AOCS (Ce 2-66; Ce 1k-07), ISO (5509) and JOCS (2.4.1.2-1996) methods. The need of a prior hydrolysis procedure was reaffirmed in a subsequent publication evaluating the AOAC method (Rozema et al., 2008). In the AOAC method the resultant FFA are converted to their FAME by addition of BF_3/methanol and subsequent heating for 40 min at 100°C.

The remaining methods recommend a saponification step by refluxing the fat in 0.5N NaOH/methanol for 5 to 15 min, followed by methylation with BF_3/methanol at refluxing conditions for 2 to 3 min (AOCS Ce2-66 & Ce 1k-07; ISO 5509), and 30 min for fish oils (ISO 5509). The ISO method 5509 offers two alternative transesterification procedures for fats and oils using trimethylsulfonium hydroxide (TMSH) as a catalyst, or 2N KOH in methanol. Each of the methods recommends direct methylation of FFA to their FAME using BF_3/methanol by refluxing the mixture for 2 to 3 min (Table 3).

- Acid digestion in the AOAC method has the advantage of hydrolyzing most known lipids, but it may cause partial or complete decomposition of some functional groups.

- The common catalyst recommended in each of these methods is BF_3, even though this catalyst has a limited shelf-life, and may result in the formation of artifacts and loss of PUFAs (Christie, 2003; Aldai et al., 2005).

- The use of BF_3 catalyst would not be appropriate to analyze the intact CLA isomeric composition produced during deodorization of vegetable oils (Juanéda et al., 2003) and during partially hydrogenation of vegetable oils (Azizian and Kramer, 2005), since this catalyst would cause further isomerization of the CLA mixture.

- Other acid-catalysts have been successfully used for the quantitative methylation of lipids, such as HCl/methanol (Stoffel et al., 1959), or H_2SO_4/methanol (Christie, 2003). These catalysts should be considered for methylation.

- The TMSH derivative has a number of limitations according to ISO method 5509. The derivatives cannot be used for cold on-column injection, they are not recommended when cyanopropyl siloxane stationary phases are used in subsequent GC analysis, and the reagent may react with hydroxyl groups. In addition, the methylation may not be quantitative for cholesteryl esters (El-Hamdy and Christie, 1993) and FFA (ISO 5509).

- The transesterification reaction using 2N KOH in methanol (ISO 5509) is a rapid and mild methylation procedure, but it does not methylate FFA, and *N*-acyl and plasmalogenic lipid if present in the sample; see discussion above.

Hydrolysis/saponification and methylation of dairy and rumen derived fats

Dairy and ruminant fats differ from plant derived fats since they contain short-chain fatty acids and acid labile CLA components that require product specific procedures. The AOAC method 996.06 recommends an alkaline hydrolysis step in which dairy products are heated with concentrated (58%) NH_4OH for 10 min at 70–80°C, prior to a subsequent methylation step using BF_3/methanol. On the

Table 3. Comparison of official methods for the preparation of FAME derivatives of plant and ruminant fats, and for GC analysis specifically for *trans* fatty acid determination.

Description	AOCS	AOCS	AOCS
Hydrolysis/saponification and methylation of plant derived fats & oils			
Official method:	Ce 2-66	Ce 2-66	
Hydrolysis/ saponification	0.5N NaOH (boil 5-10 min)	0.5N NaOH (boil 5-10 min)	
Methylation	12-15% BF$_3$ (boil 2 min)	12-15% BF$_3$ (boil 2 min)	
Transesterification 1			
Transesterification 2			
Free fatty acid	12-15% BF$_3$ (boil 2 min)	12-15% BF$_3$ (boil 2 min)	
Saponification and methylation of dairy and ruminant derived fats			
Official method:			Ce 1k-07
Saponification			0.5N NaOH (reflux 15 min)
Methylation			14% BF$_3$ (reflux 2 min)
Transesterification 1			
Transesterification 2			
Official GC methods:	Ce 1f-96	Ce 1h-05	Ce 1j-07 (RP)^c
Date approved Suitable for	Revised 2002 Vegetable oils & fats	2005 Vegetable & non-ruminant fats	2007 Dairy & ruminant fats
Not suitable for		Dairy & ruminant fats	
IS	10:0 TAG	21:0 TAG	13:0 TAG

AOAC[a]	ISO	JOCS
996.06	5509	2.4.1.2-1996
8.3M HCl	0.5N NaOH	0.5N NaOH
(70-80°C, 40 min)	(reflux 5-10 min)	(heat 5-10 min)
7% BF_3	12-15% BF_3	14% BF_3
(100°C, 45 min)	(boil 3 min)	(boil 2 min)
	0.2 mol TMSH[b] in methanol (shake vigorously for 30 s)	
	2N KOH/methanol (shake vigorously 30 s)	
996.06	5509	
NH_4OH (58%) (70-80°C, 10 min) (Dairy NH_4OH; cheese NH_4OH then HCl)		
7% BF_3 (100°C, 45 min)		
	0.2 mol TMSH[b] in methanol (shake vigorously for 30 s)	
	2N KOH/methanol (shake vigorously 30 s)	
996.06	15304	Kyo-4-1996
Revised 2001 All products	2002/2003 Vegetable oils & fats	1996/2003 Fats & oils
		Fish & milk fat
11:0 TAG	No IS, calculate relative mass	17:0 FFA or 21:0 FFA

Table 3., cont. Comparison of official methods for the preparation of FAME derivatives of plant and ruminant fats, and for GC analysis specifically for *trans* fatty acid determination.

Description	AOCS	AOCS	AOCS
Official GC methods:	Ce 1f-96	Ce 1h-05	Ce 1j-07 (RP)[c]
Column 1	SP2340 (60m)	SP2560 (100m)	SP2560 (100m)
Column 2	CP Sil 88 (50m)	CP Sil 88 (100m)	CP Sil 88 (100m)
GC temperature program	Isothermal 170°C	Isothermal 180°C	180°C - 215°C[d]
Carrier gas	He or H	H or He	H or He
Split/splitless	Split (100:1)	Split (100:1)	Split (100:1)
Injection volume	0.5-1µL	1µL	1µL
Concentration	7mg/mL	15-20mg/mL	15-20mg/mL
Chromatograms	50m & 60m separations	100m separations	100m separations
Calculations		TCF[f]	TCF[f] & ECF[g]
Conversion to TAG equivalent	g/100 g, as TAG	g/100 g, as TAG	g/100 g, as TAG
Relative to		21:0	13:0
Supporting publication	Duchateau et al., 1996	Ratnayake et al., 2008	
GC column & temp. program	Same as Ce 1f-96	Same as Ce 1h-05	
GC Chromatograms	Yes	Yes	

Abbreviations: FA, fatty acid; FFA, free fatty acid; IS, internal standard; RT, retention time; TAG, triacylglycerol; TMSH, trimethylsulfonium hydroxide
[a] See also House, S.D. (1997).
[b] TMSH; trimethylsulfonium hydroxide
[c] RP, recommended practice
[d] 180°C (32 min) - 20°C/min- 215°C (31.25 min); revised in 2009
[e] 180°C (60 min) - 10°C/min- 220°C (10 min)

AOAC[a]	ISO	JOCS
996.06	15304	Kyo-4-1996
SP2560 (100m)	SP2560 (100m)	SP2560 (100m)
	CP Sil 88 (100m)	
100°C - 240°C at 3°C/min	Isothermal 170°C or 150°C	180°C - 220°C[e]
He	H or He	He
Split (200:1)	Split (100:1)	Split (100:1)
1µL	0.5-1µL	1µL
Not specified	7mg/mL	
No GC graphs, only RT	100m separations	No GC graphs, no RT
Response factor		Relative response
g/100 g, as TAG	Expressed as relative mass	Area % of total FA composition
16:0		12:0
Rozema et al., 2008		Shirasawa et al., 2007
SP2560 (100m)		SP2560(100m)
170°C - 210°C[h]		180°C - 220°C[e]
Yes		Yes

[f] TCF, Theoretical FID response correction factor, or 1/TRF (theoretical FID response factor)

[g] ECF, Empirical correction factor determined using reference FAME mixture #714 from Nu Chek Prep

[h] 170°C (11.2min) - 7.9°C/min - 200°C (1.3min) - 7.9°C/min - 210°C (16min)

other hand, the method recommends that cheese products be treated in sequence using both NH_4OH and HCl hydrolysis, before methylation with BF_3/methanol.

The AOCS method Ce 1k-07 recommends a saponification step for dairy and ruminant fats in which samples are refluxed with 0.5N NaOH in methanol for 15 min, prior to subsequent methylation using BF_3/methanol. The ISO method 5509 recommends a direct transesterification step to form FAME using either 2M KOH/methanol or TMSH/methanol, in which mixtures are vigorously shaken for 30 s until clarified.

- The alkali treatment in each of these methods addresses the acid labile nature of CLA in dairy and ruminant fats that prevents isomerization of the CLA isomers (Kramer et al., 1997).

- Alkali hydrolysis or transesterification steps will exclude lipids stable to alkali such as sphingomyelin, plasmalogenic lipids and FFA from the total fat. In dairy, cheese and tallow these amounts are minor (about 0.3% of total fat). However, the amounts excluded are not minor in samples containing red meats, either from ruminants or non-ruminants. The red meat portion (polar lipids) contains appreciable amounts of sphingomyelin and the plasmalogenic lipids; see discussion above.

- The BF_3-catalyzed methylation procedure is a concern since it will result in isomerization of CLA, depending on the temperature and conditions used (Kramer et al., 1997; Yurawecz et al., 1999).

- There is a concern that the various methods do not take into consideration possible losses of volatile FAME in the head space and during subsequent manipulations. In general, relatively large containers are used for the preparation of FAME from dairy, cheese and ruminant products. The AOCS methods Ce 1k-07 and Ce 2-66 use 50 mL flat bottom flasks, the AOAC method 996.06 uses Mojonnier flasks (25 mL), and the ISO method 5509 uses 50 mL or 100 mL flasks.

- Each of these methods also calls for refluxing the mixture after addition of BF_3. However, the reaction times are different, 2 min for some methods (Ce 2-66, Ce 1k-07, and 2.4.1.2-1966), 3 min for ISO method 5509, and 45 min for AOAC method 996.06 and Rozema et al. (2008); the latter appears to be too long.

- Quantitation of 4:0 is a concern. The AOAC method 996.06 calls for an extraction step after hydrolysis that may lead to losses of butyric acid, since butyric acid is miscible with water. Furthermore, the resultant FAME after methylation with BF_3 are washed with an aqueous solution in both the AOAC 996.06 and AOCS Ce 1k-07 method, a step that may lead to further loss of methyl butyrate since the latter is slightly soluble in water (1.5g/100 mL). A study may need to establish total recovery and analysis of the short-chain FAME (Ce 1k-07/recommended practice Ce 1j-07 and AOAC 996.06) as proposed in ISO method 5509 for TMSH.

- There is a specific concern about AOAC method 996.06 that recommends a combination of NH_4OH and acid hydrolysis for the analysis of cheese lipids. The inclusion of an acid digestion step may be the cause for the high levels of *trans,trans*-CLA in cheese lipids as reported by some (Shantha et al., 1992 & 1995), which could not be confirmed by avoiding the acid digestion procedure (Yurawecz et al., 1999).

- An alternative transmethylation procedure for the analysis of dairy and ruminant fats should be considered that addresses the concerns of both short-chain fatty acids and CLA. The method would involve the use a Na methoxide-catalyzed methylation procedure conducted under anhydrous conditions. It was first proposed by Christie (1982), modified by for milk fat by Chouinard et al. (1999), and miniaturized for the use of 2 mL autosampler vials by Cruz-Hernandez et al., (2004). Such a method would address the concern of retaining the CLA structures and minimizing the loss of volatile fatty acids.

- All the methods include a BF_3 methylation step, and the AOAC method includes an addition HCl digestion step, which will convert plasmalogenic lipids, if present, to dimethylacetals; see discussion above. Ignoring these dimethylacetals leads to a decreased total fat content of the sample, since these long-chain aldehydes are converted to their corresponding fatty acids in biological systems and thus contribute to the total fatty acid pool (Santercole et al., 2007; Cruz-Hernandez et al. ,2006; Kraft et al., 2008; Aldai et al., 2009). Therefore, a decision will need to be made of whether to include the methoxy derivatives of the long-chain alk-1-enyl ethers formed during acid-catalyzed methylation of ruminant meat fats as part of the total fat content of the product.

- Meat lipids have been successfully analyzed using separate procedures of HCl- and $NaOCH_3$-catalyzed methylations (Cruz-Hernandez et al., 2006; Santercole et al., 2007; Kraft et al., 2008). The exclusion of plasmalogens will also lead to a lower total fat content, since the long-chain aldehydes contribute to the total fatty acid pool, and to a higher total *trans* content since some aldehydes contain *trans* double bonds (Wolff, 2002). On the other hand, sphingomyelins consisting mainly of saturated and mono-unsaturated fatty acid would hence contribute only to the total fat content of a product.

GC column selections

Currently, there is general agreement that the 100 m fused silica capillary columns coated with highly polar, 100% cyanopropyl polysiloxane stationary phases are mandatory for the *trans* fatty acid determination of all plant and ruminant fats. Each of the methods listed recommends these columns, except the earlier AOCS method Ce 1f-96, which was only included in this list for comparison. This earlier methods, like those from other organizations were based mainly on the shorter GC capillary columns, such as the 50 m CP Sil 88 column, the 50 m BPX column, the 60 m SP 2340 column, or the 60 m TC-70 column. These shorter GC columns extensively underestimated the *trans* fatty acid content (Duchateau et al., 1996; Wolff and Precht, 2002; Goley et al., 2006), and therefore, they will not be discussed here any further.

The only GC columns that meet the requirement for maximum *trans* fatty acid separation are the 100 m SP2560 (Supelco Inc.), the 100 m CP Sil 88 (Varian Inc.), and the 100 m Rtx 2560 (Restek) columns.

There are several other 100 m (or longer) columns currently available in which the cyanopropyl polysiloxane stationary phase is bonded that offer greater column stability. However, these columns are only equivalent to 70% of the cyanoproply polysiloxane columns. These columns include the 100 m HP 88 (Agilent Technologies), the 120 m BPX-70 (SGE Inc.), and the 60 m TC-70 (GL Sciences Inc.; Shirasawa et al., 2007). The separation characteristics of the *trans*- and *cis*-18:1 isomers of each of these columns does not meet the required separation demanded in the official methods.

The 200 m CP SELECT (Varian Inc.) column deserves special mention since this 200 m column effectively resolves most of the *trans*- and *cis*-18:1 isomers. As seen in Fig. 12, the separations match, and occasionally exceed the resolution obtained using the officially recognized GC columns; see further discussion above. The potential of the 200 m CP SELECT column will need to be developed further.

The choice of internal standard

Ideally, the internal standard should not be present in the matrix to be analyzed. This is potentially possible in plant derived lipid, because of the smaller number of fatty acids present in plant fats. However, this requirement is virtually impossible in dairy and ruminant fats, since dairy fats are reported to contain up to 400 different fatty acids that include acyl chains ranging from C4 to C26, with double bonds ranging from 0 to 6, and many geometric and positional isomers, as well as branched-chain-, oxo-, keto-, and hydroxy-fatty acids (Jensen, 2002). As shown in Table 3, different internal standards were selected in each of the methods, generally as TAG, but also as FFA (JOCS Kyo-4-1996). The reason for the selection of the internal standard is not provided in any of the official methods.

The choice of 21:0 for plant lipid composition and their partially hydrogenated fats and fully refined oils (Ce 1h-06 and Kyo-4-1996) in unfortunate, since this internal standard elutes in the CLA region with these columns. Both partial hydrogenation (Azizian and Kramer, 2005) and the deodorization step during the refining of oils (Juanéda et al., 2003) will produce small amounts of CLA. The analysis of the CLA isomers in these fats will be more difficult with the addition of 21:0, and the CLA isomers may interfere with the quantitation of fatty acids. The choice of 13:0 TAG would have been a better choice for plant derived lipids.

The choice of 17:0 in the JOCS method is also unfortunate since this fatty acid is ubiquitously found in all plant oils.

The choice of 13:0 TAG as an internal standard in dairy and ruminant fats is reasonable since this fatty acid is present in only minor amounts in these fats. The only interference of 13:0 may be encountered with *c*9-12:1 that is also present in ruminant fats.

The selection of GC temperature program

The search for the ideal temperature setting to analyze all FAME with the most currently used columns in a single analysis is a challenge that has yet to be met. New official methods are continuously being proposed in response to new and more efficient GC columns. Differences in the GC temperature settings between these official methods (Table 3) is evidence of the difficulty of selecting a GC program that maximizes the separation of all FAME in different fat samples. There are obvious differences in the nature of the fatty acid composition between different fats that demand product-specific methods for FAME preparation as already noted, and therefore, equally important is the need for product-specific GC temperature settings. One of the primary goals of the official methods is to provide reliable methods to meet labeling requirements. In addition, these methods must be able to determine accurate values for total saturated-, mono-unsaturated- and polyunsaturated fatty acids in foods Therefore, the GC program must meet the same standard of accuracy demanded by GC programs developed for research.

Plant derived fats and oils consist mainly of C18 fatty acids with up to 3 double bonds with minor amounts of 16:1, 20:1, 22:1 and 24:1 depending on the oil source. Partial hydrogenation and deodorization leads mainly to geometric isomerization (Wolff 1992; Precht and Molkentin, 2000c; Juanéda et al., 2003), which in the case of plant oils is limited to C18 geometric and positional isomers, with minor amounts of C16 and C20 isomers. The separation of these *trans*- and *cis*-18:1 isomers were best resolved using isothermal conditions at 175°C with these 100 m capillary columns coated with 100% cyanopropyl polysiloxane stationary phases (Molkentin and Precht, 1995). Later others reevaluated the separation of the 18:1 isomers at 170°C, 180°C and 190°C isothermal conditions, and concluded that 180°C provided the best resolution (Ratnayake, 2001; Ratnayake et al., 2002); the 180°C isothermal separation was subsequently adopted in AOCS method Ce 1h-05. Several criteria are generally set to judge good chromatographic conditions such as a separation between *t*13/*t*14-18:1 and *c*9-18:1, and base line resolution between *c*9-18:1 and *c*11-18:1, and between *c*11-20:1 and *t*9,*c*12,*c*15-18:3 (Ratnayake et al., 2002 and 2006).

- The unsaturated fatty acids in vegetable oils consist mainly of 18:1, 18:2 and 18:3 with minor amounts of 16:1 and 20:1. Partial hydrogenation and/or deodorization will produce mainly geometric and some positional isomers of these fatty acids that can be resolved using isothermal GC conditions at 180°C as specified in this method.

- The isothermal separation at 180°C resolved all *trans*-18:1 isomers, except *t*15-18:1. Resolution of the *t*13-/*t*14-18:1 isomer pair from *c*9-18:1 depends on their relative abundances; it works best with higher *trans* fats. All *cis*-18:1 isomers were resolved, except *c*6-*c*8-18:1 and *c*16-18:1, while *c*14-18:1 may occasionally overlap with the *t*16-18:1 isomer. Several *t/t*- and *c/t*-18:2, and several *c/t/t*- and *c/c/t*-18:3 isomers were resolved under these conditions, except *t*9*c*12*c*15-18:3 from either *c*9-20:1 or *c*11-20:1. The latter can be resolved by lowering the column temperature (Wolff 1994; Kramer et al., 2008).

- The scope of AOCS 1h-05 and ISO 15304 clearly indicates that these methods are not suitable for the analysis of the fats from dairy, ruminant and fish sources. This exclusion is appropriate since isothermal GC conditions at 180°C would not resolve the short-chain fatty acids in dairy and ruminant fats (Precht and Molkentin, 1996; Kramer et al., 2001; Cruz-Hernandez et al., 2004), and the BF_3 methylation procedure would isomerize some *c/t*-CLA isomers in ruminant fats to their *t/t*-CLA isomers (Kramer et al., 1997).

- Several of the test samples used to validate the official method Ce 1h-05 were determined to have *trans* levels of 1% or less of total fat. For these low-level test samples, the corresponding HORRAT parameters that are a measure of precision among laboratories, were outside the expected range of 0.5-2.0 values. As a result, the AOCS Ce 1h-05 did not specify a lower limit of quantitation.

Dairy and ruminant fats differ from plant derived fats in their complexity with over 400 different fatty acids that includes in addition to the fatty acids generally found in plant fats and oils, an abundance of short-chain fatty acids staring from 4:0, branch-chain fatty acids, long-chain PUFA with up to 6 double bonds, and an array of geometric and positional isomers (Jensen, 2002). Many chromatographic separations have been published to try to resolve as many of the fatty acids in dairy fats as possible in one analysis. Each of these separations involves some kind of temperature program to insure the resolution of the short-chain fatty acids from the solvent front, then proceed through a plateau region to resolve the complex mixture of C16, C18 and C20 isomers, and finally proceed to the maximum temperature permissible for these column to elute the long-chain PUFA and SFA. Several of the separations are listed in Table 4.

One approach to better resolve the complex 16:1, 18:1, 18:2, 18:3 and 20:1 regions of milk fats was to complement the total milk fat analysis with isothermal operation at 175°C (Molkentin and Precht, 1995). Another approach was to perform prior separation using silver-ion chromatographic techniques followed by GC analysis of the isolated fractions and then combining the results (Molkentin and Precht, 1995; Precht and Molkentin, 1996, 1997, 1999, 2000a, 2000b and 2003; Kramer et al., 2001, 2002 and 2008; Cruz-Hernandez et al., 2004 and 2006; Golay et al., 2006; Destaillats et al., 2007). This approach does resolve the different isomers, but the procedure is lengthy, and requires skill. A third approach was published recently in which the samples are analyzed two times in sequence using two separate GC temperature programs, with the same GC instrument and the same GC column (Kramer et al., 2008). Differences in column operating conditions will result in marked differences in the relative elution of different FAME groups to permit their identification; see discussion above (Fig. 14–17). This technique has been used successfully in the identification of all geometric and positional

Table 4. Typical Temperature Programs Used for Analysis of Total Milk Fat FAME

Column type	Temperature program: (temperature (hold) - rate - temperature (hold) - rate -, etc)	Total min	Reference
CP Sil 88	45 °C(1 min)-5 °C/min-225 °C(25 min)	62	Precht & Molkentin, 1996
SP2560	70 °C(4 min)-13 °C/min-175 °C(27 min)-4 °C/min-215 °C(31 min)	80	Kramer et al., 1997
CP Sil 88	75 °C(2 min)-5 °C/min-170 °C(40 min)-5 °C/min-220 °C(20 min)	91	Roach et al., 2000
CP Sil 88	45°C(4 min)-13°C/min-175°C(27 min)-4°C/min-215°C(35 min)	86	Kramer et al., 2002 & 2008
CP Sil 88	45°C(4 min)-13°C/min-163°C(37 min)-4°C/min-215°C(40 min)	103.8	Kramer et al., 2008
CP Sil 88	45°C(4 min)-13°C/min-150°C(47 min)-4°C/min-215°C(35 min)	110.3	Kramer et al., 2008
CP Sil 88	60°C(5 min)-15°C/min-165°C(1 min)-2°C/min-225°C(17 min)	60	Dionisi et al., 2002
CP Sil 88	70°C(1 min)-5°C/min-100°C(2min)-10°C/min-175°C(34min)-4°C/min-225°C (22min)	85	Shingfield et al., 2006
CP Sil88-He	80°C(1min)-25°C/min-160°C(3min)-1°C/min-190°C(5min)-2°C/min-230°C(25min)	87.5	Wąsowska et al., 2006
SP2560	170°C(4 min)-3°C/min-240°C(15 min)	42.3	AOAC 996.06
SP2560 or CP Sil 88	170°C(11.2 min)-7.9°C/min-200°C(1.3 min)-7.9°C/min-210°C(16 min)	33.6	Rozema et al., 2008
SP2560 or CP Sil 88	180°C(32 min)-20°C/min-215°C(31.25 min)	64.5	AOCS 1j-07 (Recommended practice, 2009)

isomers in the 16:1, 18:1, 18:2, 18:3 and 20:1 regions in both milk and meat fat of ruminants (Dugan et al., 2007; Kraft et al., 2008; Aldai et al., 2008 and 2009). Even though this approach is not strictly based on a single GC separation, the dual GC analysis requires no extra procedures or work up, since the same instrument and GC column are used. The only extra work involved is to match the different fatty acid isomers from both separations during data manipulation which can be automated. This method would be ideal to determine the total fatty acid composition of these fats, and provide a more accurate analysis of all the *trans*-18:1 isomers as well the *trans*-16:1 and *trans*-20:1 isomers present in milk fat that are generally overlooked. In addition, this method makes possible the analysis of the complex *cis/trans*-18:2 isomers and SFA coeluting in this region (Kramer et al., 2008).

- Proposing a GC program for milk fats that starts at 180°C followed by a ramp to elute the long-chain PUFA and SFA fails to adequately address the separation of short-chain fatty acids. It is for this reason, most researchers in this field have used lower starting temperatures to analyze the short-chain FAME in dairy fats (Table 4).

- The criteria of a separation between *t*13/*t*14-18:1 and *c*9-18:1 in dairy and ruminant fats is difficult to achieve unless specific isothermal conditions are selected (Molkentin and Precht, 1995) that may not be suitable for the remaining fatty acids in the mixture. The resolution of *t*15-18:1 from c9-18:1 is only achieved at lower column temperatures of 163°C (Kramer et al., 2008). The resolution of the *c/t*-18:2 from saturated fatty acids eluting in this region requires analyses at different temperatures (Precht and Molkentin, 2003). The resolution of *trans*- from the *cis*-20:1 isomers, and the coeluting *c/c/t*-18:3 isomers requires different operating temperatures (Wolff, 1994; Kramer et al., 2008).

- The approach to use two separate GC programs to analyze dairy and ruminant fats should be considered because it addresses the need for accurately determining the short-chain FAME, preserving the CLA isomer profile, and resolving the complex mixture of *trans* fatty acids in dairy and ruminant fats.

- Regardless of which GC program is recommended, additional silver-ion HPLC separations are required to analyze several CLA isomers since they coelute in all GC separations, i.e., *c*9*t*11-19:2 and *t*7*c*9-18:2; see further discussion above (Fig. 22–25).

Other GC parameters

In general, the GC parameters in the different official methods call for consistency. The recommendation to use hydrogen as carrier gas is commendable. The type of injection system may need to be reassessed to minimize discrimination (Grob and Biedermann, 2002), particularly regarding dairy and ruminant fats that contain both short- and long-chain FAME. This is less of a problem with plant oils since the range of FAME is much smaller.

Calculations

Theoretical correction of FID responses is an absolute necessity. The use of standard FAME mixtures to determine empirical correction factors using mixtures with known composition, such as #714 from Nu Chek Prep, is an excellent approach, but care is required to ensure the accurate of each component in this and other mixtures. There may be a need to work with suppliers to assess the purity of each of the FAME in the mixture. The certificate of analysis is only as good as the accuracy of its components; see discussion above.

Fourier-Transform Infrared Spectroscopy

Instrumentation

A brief introduction of today's infrared technology will help the reader understand the *trans* fat methodologies reviewed below. A Fourier transform infrared (FTIR) spectrometer (Reedy and Mossoba, 1999) consists of a source of continuous infrared radiation that emits light from a high temperature element that withstands prolonged heating and exposure to air, an interferometer, and a detector. The interferometer allows the detection of all the component wavelengths of the mid-infrared region (4000–600 cm^{-1}) simultaneously. When a test sample (such as a *trans* fat solution) is placed between the beam splitter and the detector, it selectively absorbs infrared energy. Changes in the energy reaching the detector as a function of time yield an interferogram, the raw infrared spectrum. When the interferogram is converted from the time to the frequency domain by the mathematical Fourier *trans*formation, a single-beam spectrum (Fig. 26) is obtained. The single-beam spectrum of a "test sample" is the emittance profile of the infrared source as well as the absorption bands of all infrared-absorbing material in the path of the infrared beam, namely the test sample, atmospheric water vapor, and CO_2. A "background" (such as a solvent or a *trans*-free fat solution) single-beam spectrum is also measured. To observe the conventional transmission (or absorption) spectrum of a test sample (Fig. 27), the single-beam spectrum of the test sample is digitally "ratioed" against the single-beam spectrum of the reference background.

FTIR instrumentation offers several advantages over dispersive spectrometers that use prisms or diffraction gratings to resolve the infrared light into its component wavelengths (Reedy and Mossoba 1999). An entire FTIR spectrum can be measured in a single scan in about 1 sec. A high signal-to-noise ratio can be typically achieved in 1–3 min.

Wavelength precision is provided by an internal reference laser. The computing capabilities offer powerful data-handling and quantitative manipulation routines. The higher energy throughput allows the efficient use of different sample handling techniques, such as attenuated total reflection (ATR) (Harrick, 1967; Mirabella, 1992; Ismail et al., 1998).

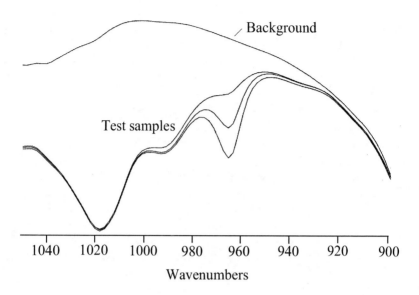

Fig. 26. Single-beam spectra for carbon disulfide (CS_2) solvent (background) and CS_2 solutions of *trans* fat (test samples).

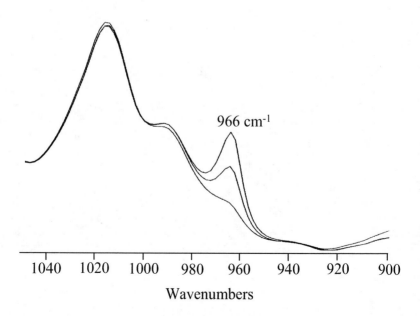

Fig. 27. Absorption spectra for CS_2 solutions of *trans* fat (test samples).

Transmission mode

In the standard transmission mode (Reedy and Mossoba, 1999; Ismail et al., 1998), certain frequencies are absorbed as the infrared beam passes through the test sample, and only the transmitted light reaches the detector and is measured. *Trans*mission liquid cells have traditionally been used to determine *trans* fats. They are made of a pair of salt crystals (such as NaCl) that are separated by a thin Teflon spacer. As indicated by Beer's law, A = a · b · c, the absorbance (A) depends on the molar absorption coefficient at a particular wavelength (a), the path length of infrared light through the test sample (b) that is dictated by the thickness of the spacer (up to 1.00 mm), and the concentration of the absorbing analyte (c). To determine the amount of an unknown, (a) is first calculated by generating a plot of the absorbance of calibration standards (e.g., carbon disulfide (CS_2) solutions of methyl elaidate) at different concentrations over the range of interest. When the analyte is a neat fat or oil (without solvent), the thickness of the spacer should be below 10 μm for the transmitted infrared radiation to reach the detector. This very short pathlength limitation is easily met by using ATR sampling techniques (Harrick, 1967; Mirabella, 1992; Ismail et al., 1999).

Attenuated total reflection mode

When melted fat or oil is placed on the surface of a crystal such as diamond, the infrared light penetrates a distance of only a few μm into the test sample when the conditions of total internal reflection apply (Harrick, 1967; Mirabella, 1992; Ismail et al., 1998). These conditions occur when light traveling in a transparent medium of high refractive index (η_1) (such as diamond or ZnSe) strikes the interface between this medium and another transparent medium of lower refractive index (η_2) (such as air or melted *trans* fat) at an angle of incidence (θ) exceeding the critical angle (θ_c) defined by:

$$\theta_c = \sin^{-1}(\eta_2/\eta_1)$$

Normally, light is partially transmitted and partially reflected. However, under these conditions, it is not transmitted, but totally reflected inside the crystal (Fig. 28).

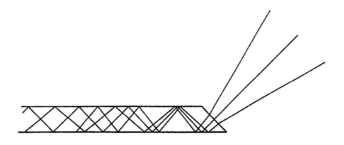

Fig. 28. Infrared light bouncing inside an internal reflection crystal.

Moreover, as the light bounces (one or more times) inside the crystal, a so-called evanescent wave also propagates away from the surface of the crystal through the melted *trans* fat (Adam et al., 2000; Mossoba et al., 2001a), in this example. At the surface of the crystal, the intensity of this wave decays exponentially with distance. It is also attenuated by the absorption of infrared light by the melted *trans* fat. The depth of penetration (d_p) of the infrared light into the test sample is minuscule. It typically varies between 1–4 μm and depends on θ, η_2, η_1, and the wavelength (λ) as given by the relation:

$$dp = \lambda/2\,\pi\eta_1\,[\sin^2(\theta) - (\eta_2/\eta_1)]^{1/2}$$

As a result, the depth of penetration will be higher the greater λ or the smaller the frequency. Therefore, an interferogram (raw infrared spectrum) is a measure of the attenuation by a *trans* fat test sample of the totally internally reflected infrared light. The interferogram of a reference background material (e.g., a *trans*-free fat) is similarly measured. They are subsequently used to obtain an absorption spectrum as explained above. ATR-FTIR measurements are easy, convenient, and require about 2 min per test sample (Adam et al., 2000; Mossoba et al., 2001a).

Trans *Fat Infrared Methodology*

Because of the interest in accurate and rapid analytical methods for quantifying total *trans* fatty acids with isolated double bonds, many infrared spectroscopic procedures and official methods have been published over the past several decades (Mossoba et al., 1996, 2003, and 2005). These proposed procedures and official methods that have been validated through national and/or international multi-laboratory collaborative studies provide varying degrees of accuracy and reproducibility. A review is given below following an introduction about the scope of this infrared determination.

The determination of total *trans* fatty acids by the different IR spectroscopic procedures (Mossoba et al., 1996) and official methods (AOCS, *Cd 14d-95*; AOAC International, *965.34*; AOCS, *Cd 14d-99*; AOAC International, *2000.10*) is based on the CH out-of-plane deformation band observed at 966 cm^{-1} (Fig. 27) that is uniquely characteristic of isolated double bonds with *trans* configuration. These double bonds are found primarily in *trans*-monoenes, and usually at much lower levels in minor hydrogenation products, such as methylene-interrupted (e.g., *trans* 9,*trans* 12-18:2) and nonmethylene-interrupted (e.g., *trans* 9,*trans* 13-18:2) *trans*,*trans*-dienes, mono*trans*dienes (e.g., *trans* 9,*cis* 12-18:2), and other *trans*-polyenes. This IR methodology has been extensively used in the fats and oils industry and found to be extremely useful to determine the triacylglycerols or fatty acid methyl esters. However, samples consisting of free fatty acids must be first esterified particularly at low *trans* levels (less than 15%) (Firestone and Sheppard, 1992) because the band near 935 cm^{-1}, due to the O-H out-of plane deformation in –C(O)OH moieties would interfere with the determination of the *trans* band at 966 cm^{-1}.

When the total isolated *trans* fat levels are relatively low (below 10%), a potentially significant interference is found in products containing approximately 1% (such as milk fat) or more of conjugated unsaturation (Firestone and Sheppard, 1992; Mossoba et al., 2001a; Firestone and LaBouliere, 1965). This is due to the fact that conjugated *trans,trans* (near 990 cm^{-1}) and/or *cis/trans* (near 990 and 950 cm^{-1}) double bonds exhibit absorption bands that are sufficiently close to, and thus interfere with, the 966 cm^{-1} band. An analytical solution to this problem based on standard addition has been published (Mossoba et al., 2001a). This procedure was recently used to determine the *trans* content of milkfat in the presence of interfering CLA isomers (Mossoba et al., 2001a).

The highly characteristic *trans* absorption at 966 cm^{-1} occurs on an elevated and sloping baseline, thus the measurement of its height or area becomes increasingly less accurate as the *trans* levels decrease (Fig. 27). Since the early report by Firestone and LaBouliere (AOAC International, 1994) that the IR determination of *trans* unsaturation yielded a high bias for triacylglycerols and a low bias for fatty acid methyl esters, many modifications have been proposed. Transmission (AOCS, *Cd 14d-95*; AOAC International, *965.34*) and internal reflection (AOCS, *Cd 14d-99*; AOAC International, *2000.10*) FTIR official methods succeeded in improving the accuracy of this determination.

Transmission FTIR Official Method AOCS Cd 14-95/AOAC 965.34

The latest transmission infrared official methods (AOCS, *Cd 14d-95*; AOAC International, *965.34*) were improvements of older ones. It requires the analysis of all samples as fatty acid methyl esters (FAME) irrespective of *trans* level. FAME test samples are accurately weighed and dissolved in known volumes of CS_2. FAME solutions are then measured by FTIR in 1-mm fixed-pathlength non-demountable transmission cells. FAME calibration standards each consisting of a known mixture of methyl elaidate and methyl oleate in CS_2 were prepared, such that the total concentration of FAMEs was the same (0.2 g/10 mL) for all standards. The total concentration of test samples was also set at 0.2 g/10 mL. This method assumes that the major component to be determined in test samples is methyl elaidate. Two linear regression calibration equations were generated, one for the set of standards with *trans* contents of ≤10% and another for those with *trans* levels >10%. To measure the 966-cm^{-1} band height, a straight line was first drawn between two points along the sloping baseline of the infrared spectrum (Fig. 27). The positions of these two points were determined by the analyst, and had to be moved closer to each other as the size of the *trans* band decreased. Validation data generated in multi-laboratory studies using this and other infrared official methods will be compared below.

ATR-FTIR Official Method AOCS Cd 14d-99/AOAC 2000.10

Using internal reflection, also known as attenuated total reflection (ATR), another official method (AOCS, *Cd 14d-99*; AOAC International, *2000.10*) was recently developed to rapidly (5 min) measure the 966-cm^{-1} *trans* band as a symmetric feature on a horizontal baseline. The experimental aspects of this ATR infrared official method are far less complex than those involving transmission measurements. This approach uses "ratioing" the *trans* test sample single-beam spectrum against that of a reference material consisting of a *trans*-free oil and applies the ATR sampling technique (Harrick, 1967; Mirabella, 1992; Ismail et al., 1998; Adam et al., 2000; Mossoba et al., 2001a) to melted fats; this avoids the weighing of test portions and their quantitative dilution with the volatile CS_2 solvent.

In today's FTIR instruments (Reedy and Mossoba 1999), single beam spectra are measured separately for both a test sample and an appropriate reference background material, and then "ratioed" to obtain an absorption spectrum. Traditionally, CS_2 solvent has been used as the reference background material in the vast majority of procedures and official methods (Mossaba et al., 1996; AOCS, *Cd 14d-95*; AOAC Interntional, *965.34*). When a *trans*-free fat reference background material is used (Fig. 29) instead of CS_2, the sloping baseline of the 966 cm^{-1} *trans* band (Fig. 27) becomes horizontal (Fig. 30) (Adam et al., 2000; Mossoba et al., 2001; Mossoba et al., 1996). Therefore, the contributions of the triacylglycerol absorptions that led to the sloping baseline in the first place are removed, and the requirement to convert triacylglycerols to FAME is eliminated.

Having a horizontal baseline minimizes the uncertainty in the measurement of the 966 cm^{-1} *trans* band area at all *trans* levels, and improves both precision and accuracy. This approach of using a *trans*-free reference background material can be used in both transmission and internal reflection modes. In transmission mode, a CS_2 solution of the test sample is "ratioed" against a CS_2 solution of a *trans*-free reference background material, or simply a neat (without solvent) test sample is "ratioed" against a neat *trans*-free reference background material. The latter is easily and rapidly achieved with an ATR cell as described by the official method and explained next.

Accurately weighed *trans* standards are prepared by adding varying amounts (0–50%) of neat trielaidin to a neat *trans*-free reference oil. Next, a small volume of a standard is placed on top of the heated (65°C) horizontal surface of the internal reflection element (usually zinc selenide or diamond) of a single-bounce ATR cell. Depending on the size of the internal reflection element, this small volume can range from 50 μL to as little as 1 μL. The element surface must be completely covered. Single beam spectra (Fig. 29) of *trans* standards (test samples) are measured by FTIR and "ratioed" against the single beam spectrum of the same *trans* free reference oil (background) to obtain absorption spectra. These spectra should exhibit the 966cm^{-1} *trans* band as a symmetric feature on a horizontal baseline (Fig. 30). The areas of the *trans* bands can then be integrated electronically between 990 and 945 cm^{-1} and used to generate a calibration curve. The resulting linear regression equation relating the integrated area and the *trans* level (as percent of total fat) of the standards usually has a negligible y-intercept and a regression coefficient R value of 0.999 (Adam et al., 2000; Mossoba et al., 2001a; AOCS, *Cd 14d-99*, AOAC International, *2000.10*).

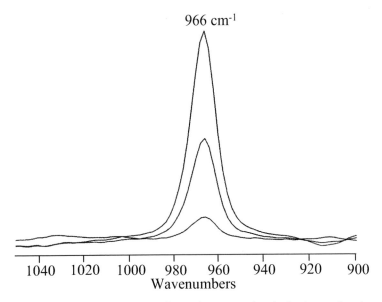

Fig. 29. Single-beam spectra for neat (without solvent) *trans*-free fat (background) and *trans* fat (test samples).

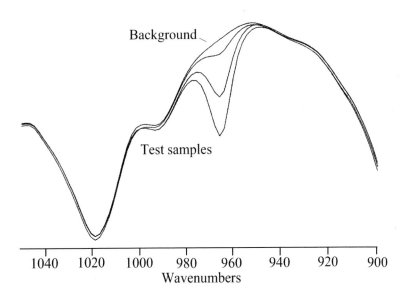

Fig. 30. Absorption spectra for neat (without solvent) *trans* fat (test samples).

Similarly, single beam spectra of unknown *trans* test samples are measured and "ratioed" against the single beam spectrum of the same *trans*-free reference oil used for calibration. The *trans* level (as percent of total fat) is then calculated by substituting the value of the integrated area of the *trans* band in the linear regression equation. This method also assumes that the major component to be determined in test samples is trielaidin.

The fatty acid composition of the *trans*-free reference oil plays a critical role. If the selected *trans*-free oil is significantly different from the matrix of the fat investigated, it may have an adverse impact on accuracy, particularly near the official method's lowest *trans* level of quantitation, 5%, as percent of total fat. This *trans*-free reference oil must be carefully selected and should resemble as much as possible the composition of the unknown *trans* fat or oil being determined.

When the corresponding FAME were used instead of triacylglycerols, similar *trans* values were obtained (Mossoba et al., 1996), thus demonstrating that this "ratioing" ATR FTIR method adequately compensates for the triacylglycerol absorptions that overlap with the 966-cm^{-1} *trans* band and contribute to its elevated and sloping baseline. In comparative studies, lower reproducibility relative standard deviation, RSD(R), values (Adam, 2000; Mossoba, 2001a) were obtained by using the ATR-FTIR method (AOCS, *Cd 14d-99*, AOAC International, *2000.10*) relative to two transmission FTIR methods (AOCS, *Cd 14d-95*; AOAC International, *965.34*) (Fig. 31).

The baseline offset and slope were eliminated by the ATR-FTIR official methods (AOCS Cd 14d-99 and AOAC 2000.10), but were only partly successful in improving accuracy. This is because it is impossible to find a reference fat that is absolutely *trans*-free and whose composition would closely match every unknown test sample. Sensitivity was also negatively impacted.

ATR-FTIR Negative Second Derivative Methodology

A novel ATR-FTIR procedure that measures the height of the negative second derivative of the *trans* absorption band (relative to an open beam) (Figs. 32–34) was recently reported to improve sensitivity and accuracy (Milosevic et al., 2004, Mossoba et al, 2007ab, 2009). The negative second derivative procedure eliminated both the baseline offset and slope of the *trans* IR band (966 cm^{-1}) as well as the requirement to use a *trans*-free reference fat. This is because the second derivative of an absorbance spectrum led to narrower bandwidths, and made it possible to notice small differences in band position. For the first time, the presence of weak interference IR bands due to saturated fat (near or below 960 cm^{-1}) were confirmed (Mossoba et al, 2007). This finding made it possible to recognize that for a fully hydrogenated soybean oil, a weak band observed at 960 cm^{-1} should be attributed to tristearin (18:0), a saturated rather than a *trans* fat, and hence allowed the resolution of a discrepancy in accuracy between GC and IR at low *trans* levels (Mossoba et al, 2007). This procedure is currently being validated in an international ATR-FTIR collaborative study.

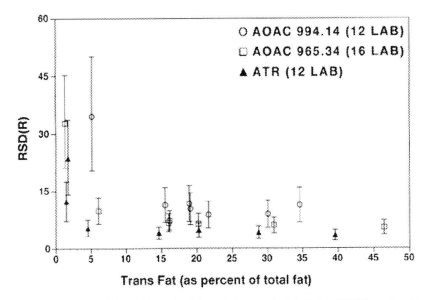

Fig. 31. Comparison of plots of reproducibility relative standard deviation, RSD(R), against the *trans* content mean values determined by three official methods: the two transmission methods AOAC 965.34 and AOAC 994.14, and the ATR method AOCS Cd 14d-99. The number of laboratories in the corresponding collaborative studies were 12,16, and 12, respectively. The error bars denote the upper and lower 95% confidence limits on the true RSD(R). (Reproduced by permission from AOCS Press).

Therefore, for fats and oils with a high content of saturated fats and only a trace amount (≤0.1% of total fat) of *trans* fat, such as coconut oil and cocoa butter, the weak bands observed at wavenumbers slightly lower than 966 cm^{-1} (Fig. 34) must not be erroneously reported as *trans* bands. This recognition of potential interferences from saturated fats would result in the correct interpretation of IR spectra for unknown *trans* fats and oils, and improve the accuracy of the IR determination at low *trans* levels (at approximately 1–5% of total fat).

In order to meet the *trans* fat labeling requirements, in particular the claim of zero gram *trans* fat per serving, it was necessary to develop a methodology with sensitivity that is capable of measuring 0.5g *trans* fat per serving. A recently published compositional database on the *trans* fat content of a wide range of foods (Satchithanandam et al., 2004) was used to estimate the *trans* fat content, as percent of total fat. For a food product that consists mostly of fat or oil, such as Ranch dressing with 27.6 g total fat per 30 g serving size (total fat 92%) (Satchithanandam et al., 2004), it is possible to calculate the corresponding value for the *trans* fat content, as percent of total fat; this value would be (0.5 g/27.6 g) × 100 = 1.8% *trans* fat (as percent of total fat). This is the minimum level of *trans* fat in a product that had to be measured with confidence to meet the declaration requirement of

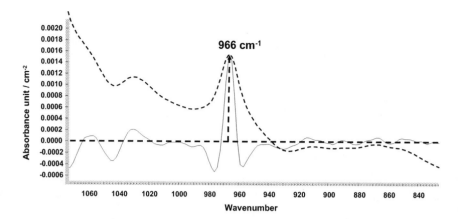

Fig. 32. Expanded region of the IR spectra that exhibit the deformation band for isolated *trans* double bonds at 966 cm⁻¹ for both the negative second derivative (solid line) and absorption spectra (dashed line) for a partially hydrogenated vegetable oil test sample. The height of the negative second derivative band (vertical arrow), can be accurately measured from a horizontal baseline. (Reproduced by permission from AOCS Press).

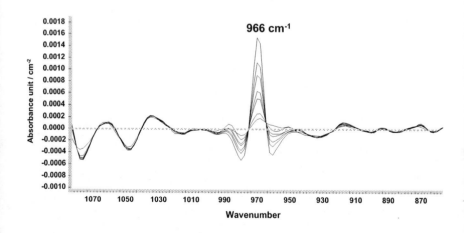

Fig. 33. Expanded spectral region that exhibits the -2nd derivative of the deformation band for isolated *trans* double bonds at 966 cm⁻¹ for representative test samples covering approximately the 1-12% range and containing *trans* fat levels determined by IR to be for Lard 1.21%, and Canola oil mixtures 2.21%; 4.20%; 4.72%; 7.35%; 9.11%; and 12.62%, as percent of total fat. The height of the -2nd derivative band can be easily measured from the horizontal baseline (dotted line). (Reproduced by permission from AOCS Press).

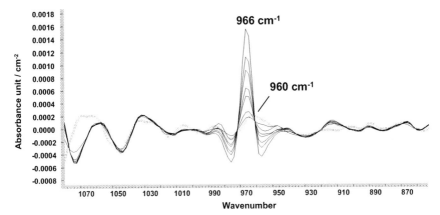

Fig. 34. Expanded spectral region that exhibits the -2nd derivative of the deformation band for isolated *trans* double bonds at 966 cm⁻¹ for several test samples containing *trans* fat (solid lines), as well as for coconut oil (dotted line) that is high in saturated fat and contains only a trace (approximately 0.1%, AOCS Ce 1h-05) of *trans* fat. Coconut oil exhibited a spectral feature at slightly lower wavenumbers, near 960 cm⁻¹, which is easy to misidentify as a band for isolated *trans* double bonds (Mossoba et al., 2007a). (Reproduced by permission from AOCS Press).

zero *trans* fat on the Nutrition Fact label. For food products containing less total fat, the level of *trans* fat in the total fat of the product would be higher than 1.8% and hence easier to determine; for example, for food products with a low total fat content per serving, typically 3 g per 28 g serving size (total fat 11%), a *trans* fat level of 0.5g per serving corresponds to a total *trans* fat content of $(0.5\text{ g}/3\text{ g}) \times 100 = 17\%$, as percent of total fat.

Upon validation of negative second derivative procedure, its *trans* fat lower limit of quantitation will be determined and is expected to be found near 1% of total fat. This relatively sensitive IR procedure should be suitable for the rapid determination of total *trans* fats in food products and dietary supplements and compliance with the food labeling regulations.

Present Status and Limitations of Official Methods

The determination of fatty acid composition and total *trans* fat by a single GC run has been the industry standard as evidenced by the various GC official methods adopted by the different international scientific associations (AOAC, AOCS, ISO, JOCS). These determinations became easier with the availability of 100 m highly polar capillary columns (Molkentin and Precht, 1995; Wolff et al., 1998; Precht and Molkentin, 1999, 2000c & 2003; Buchgraber and Ulberth, 2001; Kramer et al., 2001, 2002 & 2008; Ratnayake et al., 2002 & 2004; Cruz-Hernandez et al., 2004; Golay et al., 2006). However, this task still requires extensive expertise in identifying all possible *trans*-containing fatty acids and their isomers, many of which remain unresolved and overlap with other fatty acids (Precht and Molkentin,

1999, 2000a, 2002c & 2003; Cruz-Hernandez et al., 2004 & 2006; Ratnayake et al. 2006; Kramer et al., 2008). The improved columns no longer underestimate the total *trans* content by as much as 35% as observed previously with the use of the shorter capillary columns (Wolff and Precht, 2002; Golay et al., 2006). Accurate determinations often requires prior separation using silver-ion chromatographic techniques (TLC, HPLC or SPE) to resolve the *trans* from the *cis* geometric isomers (Henninger and Ulberth, 1994; Wolff and Bayard, 1995; Precht and Molkentin 1999, 2000c & 200a; Kramer et al., 2001 & 2008; Cruz-Hernandez et al., 2004 & 2006) followed by GC analysis of the isolated fractions; at much lower GC isothermal conditions all individual *trans* fatty acid isomers are resolved (Precht and Molkentin, 1999, 2000a, & 2000c; Cruz-Hernandez et al., 2004 & 2006). The ideal chromatographic methodology for the quantitation of all *trans* fatty acids has yet to be developed and validated. GC official methods do not state lower levels of quantitation, but total *trans* levels as low as 0.06%, as percent of total fat have been reported (AOCS Ce 1h-05).

Infrared spectroscopy has been the tool of choice for the rapid determination of total isolated *trans* double bonds in oils and fats (AOCS, Cd 14-95; AOAC, 965.06). Significant improvements were introduced by the new ATR-FTIR official methods (AOCS, Cd 14d-99; AOAC, 2000.10). However, the latest FTIR official method exhibited poor precision for *trans* levels below 5%, as percent of total fat (AOAC, 2000.10). To overcome the limitations responsible for such outcome, the so-called negative second derivative ATR-FTIR was recently proposed (Milosevic et al., 2004, Mossoba, et al., 2007ab, & 2009) and is currently being validated in an international collaborative study. By measuring the second derivative, instead of the *trans* absorption band itself, spectral features were enhanced such that *trans* fatty acid levels as low as approximately 1% could be readily measured (Milosevic et al., 2004, Mossoba, et al., 2007ab, & 2009). This methodology can be applied conveniently to the determination of the total *trans* fat content in food products. The FTIR methods are more rapid than GC methods, but do not provide detailed information on fatty acid composition. Although for *trans* fatty acid nutrition labeling, information regarding specific *trans* isomeric composition is currently not required, except for the exclusion of CLA (DHHS, 2003). Differentiation of the *trans* absorption band attributed to CLA isomers containing a conjugated *trans* double bond from that of isolated *trans* fatty acids had been a challenge (Kramer et al., 1997). However, resolution of these IR absorption bands by using second derivative FTIR spectra with narrower bandwidths (Milosevic et al., 2004, Mossoba et al, 2007 and 2009) could now be achieved.

There is a specific need to address differences in total *trans* fatty acid content that are obtained by chromatographic and spectroscopic techniques. This problem may perhaps be related to unidentified *trans* fatty acids by GC peaks, non-acyl lipid like alkenyl ethers that also contain *trans* double bonds (Wolff 2002), non-*trans* fat interference infrared absorption bands, or to the nature of specific food matrices. Consistencies between GC and FTIR methods will need to be validated for accuracy, reliability and applicability to different foods.

Recent advances in Fourier *trans*form-near infrared (FT-NIR) have made it possible to determine not only the total *trans* fatty acid content of a fat or oil as it is done by ATR-FTIR, but also its fatty acid composition (Azizian and Kramer, 2005), and without the need for prior derivatization to volatile derivatives as required for GC analysis. Quantitative FT-NIR models were developed by comparing accurate GC results with FT-NIR measurements and by using chemometric analyses. FT-NIR has also been applied to the determination of CLA in the presence of *trans* fatty acids (Christy et al., 2003). FT-NIR methodologies show great potential, but they will need to be validated in collaborative studies.

At the present time, the *trans* fat regulations in the US, Canada and many other countries require neither a distinction between the different *trans* fatty acid isomers, nor between *trans*-containing mono-, di- and tri-unsaturated fatty acids. However, there will be a need to separate and possibly report the various *trans* fatty acid isomers, as evidence mounts that certain *trans* isomers have nutritionally and/or physiologically beneficial properties, while others have undesirable effects. For example, the *t*11-18:1 isomer was shown to be a precursor for the conjugated *c*9,*t*11-18:2 isomer in mammalian tissues (Griinari et al., 2000; Bauman and Griinari, 2001), and the latter CLA isomer has reportedly beneficial physiological effects in laboratory animals (Pariza et al., 2001; Belury 2002; Martin and Valeille 2002) and humans (Aro et al., 2000; Tricon et al., 2004; Rajakangas et al., 2003). To report individual *trans* fatty acids, the GC method will become an essential research tool to identify beneficial isolated and conjugated *trans* isomers. However, the current *trans* fat nutrition labeling regulation requires only the declaration of the **total** *trans* content for food products and dietary supplements. This requirement could certainly be met more conveniently and rapidly by using ATR-FTIR spectroscopy.

References

Ackman, R.G.; S.N. Hooper; D.L. Hooper. Linolenic acid artifacts from the deodorization of oils. *J Am Oil Chem Soc,* **1974,** *51,* 42–49.

Ackman, R.G.; T.K. Mag. *Trans* fatty acids and the potential for less in technical products. In: Sébédio J-L, Christie WW (eds), *Trans* Fatty Acids in Human Nutrition, The Oily Press, Dundee, Scotland, 1998, pp. 35–58.

Adam, M.; M.M. Mossoba; T. Lee. Rapid determination of total *trans* fat content by attenuated total reflection infrared spectroscopy: An international collaborative study. *J Am Oil Chem Soc,* **2000,** *77,* 457–462.

Aldai, N.; M.E.R. Dugan; J.K.G. Kramer; P.S. Mir; T.A. McAllister. Non-ionophore antibiotics do not affect the *trans*-18:1 and CLA composition in beef adipose tissue. *J Anim Sci,* **2008,** *86,* 3522–3532.

Aldai, N.; M.E.R. Dugan; D.C. Rolland; J.K.G. Kramer. Survey of the fatty acid composition of Canadian beef: 1. Backfat and *longissimus lumborum* muscle. *Can J Anim Sci,* **2009,** (in press).

Aldai, N.; B.E. Murray; A.I. Nájera; D.J. Troy; K. Osoro. Review. Derivatization of fatty acids and its application for conjugated linoleic acid studies in ruminant meat lipids. *J Sci*

Food Agric, **2005**, *85*, 1073–1083.

Aitchison, J.M.; W.L. Dunkley; N.L. Canolty; L.M. Smith. Influence of diet on *trans* fatty acids in human milk. *Am J Clin Nutr,* **1977**, *30*, 2006–2015.

AOAC Official Method 994.14, Isolated *Trans* Unsaturated Fatty Acid in Partially Hydrogenated Fats. 1994.

AOAC Official Method 965.34, Isolated *Trans* Isomers in Margarine and Shortenings. 1997.

AOAC Offical Method 996.06, Fat (Total, Saturated, and Unsaturated) in Foods. Hydrolytic Extraction Gas Chromatographic Method. Revised 2001.

AOAC Official Method 2000.10, Determination of Total Isolated *trans* Unsaturated Fatty Acids in Fats and Oils. ATR-FTIR Spectroscopy. 2000.

AOCS Official Method Ce 1f-96, Determination of *cis-* and *trans-* Fatty Acids in Hydrogenated and Refined Oils and Fats by Capillary GLC. Revised 2002.

AOCS Official Method Ce 1h-05, Determination of *cis-*, *trans-*, Saturated, Monounsaturated and Polyunsaturated Fatty Acids in Vegetable or Non-ruminant Animal Oils and Fats by Capillary GLC. Approved 2005.

AOCS Recommended Practice Ce 1j-07, Determination of *cis-*, *trans-*, Saturated, Monounsaturated, and Polyunsaturated Fatty Acids in Dairy and Ruminant Fats by Capillary GLC. New 2007.

AOCS Official Method Ce 1k-07, Direct Methylation of Lipids for theDetermination of Total Fat, Saturated, *cis-*Monounsaturated, *cis-*Polyunsaturated, and *trans* Fatty Acids by Chromatography. Revised 2007.

AOCS Official Method Ce 2-66, Preparation of Methyl Esters of Fatty Acids. Reapproved 1997.

AOCS Official Method Cd 14d-95, Isolated *trans* Isomers Infrared Spectroscopic Method. Reapproved 1997.

AOCS Official Method Cd 14d-99, Rapid Determination of Isolated *trans* Geometric Isomers in Fats and Oils by Attenuated Total Reflection Infrared Spectroscopy. Revised 1999. *965.342000.10*

Aro, A.; S. Männistö; I. Salminen; M.-L. Ovaskainen; V. Kataja; M. Uusitupa. Inverse association between dietary and serum conjugated linoleic acid and risk of breast cancer in postmenopausal women. *Nutr Cancer,* **2000**, *38*, 151–157.

Azizian, H.; J.K.G. Kramer. A rapid method for the quantification of fatty acids in fats and oils with emphasis on *trans* fatty acids using Fourier Transform Near Infrared Spectroscopy (FT-NIR). *Lipids,* **2005**, *40*, 855–867.

Bauman, D.E.; J.M. Griinari. Regulation and nutritional manipulation of milk fat: low-fat milk syndrome. *Livest Prod Sci,* **2001**, *70*, 15–29.

Bauman, D.E.; J.M. Griinari. Nutritional regulation of milk fat synthesis. *Annu Rev Nutr,* **2003**, *23*, 203–227.

Bauman, D.E.; D.M. Barbano; D.A. Dwyer; J.M. Griinari. Technical note: Production of butter with enhanced conjugated linoleic acid for use in biomedical studies with animal models. *J Dairy Sci* **2000**, *83*, 2422–2425.

Bell, J.A.; J.M. Griinari; J.J. Kennelly. Effect of safflower oil, flaxseed oil, monensin, and vitamin E on concentration of conjugated linoleic acid in bovine milk fat. *J Dairy Sci*, **2006**, *89*, 733–748.

Belury, M.A. Dietary conjugated linoleic acid in health: physiological effects and mechanisms of action. *Annu Rev Nutr*, **2002**, *22*, 505–531.

Bessa, R.J.B.; P.V. Portugal; I.A. Mendes; J. Santos-Silva. Effect of lipid supplementation on growth performance, carcass and meat quality and fatty acid composition of intramuscular lipids of lambs fed dehydrated lucerne or concentrate. *Livest Prod Sci*, **2005**, *96*, 185–194.

Buchgraber, M.; F. Ulberth. Determination of *trans* octadecenoic acids by silver-ion chromatography-gas liquid chromatography: An intercomparison of methods, *J AOAC Internat*, **2001**, *84*, 1490–1498.

Buchgraber M.; F. Ulberth. Determination of low level *trans* unsaturation in physically refined vegetable oils by capillary GLC - Results of 3 intercomparison studies. *Eur J Lipid Sci Technol*, **2002**, *104*, 792–799.

Castello, G.; S. Vezzani; G. D'Amato. Effect of temperature on the polarity of some stationary phases for gas chromatography. *J Chromatogr A*, **1997**, *779*, 275–286.

Chardigny, J.-M.; R.L. Wolff; E. Mager; C.C. Bayard; J.-L. Sébédio; L. Martine; W.M.N. Ratnayake. Fatty acid composition of French infant formulas with emphasis on the content and detailed profile of *trans* fatty acids. *J Am Oil Chem Soc*, **1996**, *73*, 1595–1601.

Chen, Z.Y.; G. Pelletier; R. Hollywood; W.M.N. Ratnayake. *Trans* fatty acid isomers in Canadian human milk. *Lipids*, **1995**, *30*,15–21.

Christie, W.W. A simple procedure for rapid transmethylation of glycerolipids and cholesterol esters. *J Lipid Res*, **1982**, *23*, 1072–1075.

Christie, W.W. Lipid Analysis: Isolation, Separation, Identification and Structural Analysis of Lipids. Third edition, The Oily Press, PJ Barnes & Associates, Bridgewater, England, 2003; p 208.

Christy, A.A.; P.K. Egeberg; E.T. Ostensen. Simultaneous quantitative determination of isolated *trans* fatty acids and conjugated linoleic acids in oils and fats by chemometric analysis of the infrared profiles. *Vibrational Spectroscopy*, **2003**, *33*, 37–48.

Chouinard P.Y.; L. Corneau; D.M. Barbano; L.E. Metzger; D.E. Bauman. Conjugated linoleic acids alter milk fatty acid composition and inhibit milk fat secretion in dairy cows. *J Nutr*, **1999**, *129*, 1579–1584.

Collomb M.; U. Bütikofer; R. Sieber; J.O. Bosset; B. Jeangros. Conjugated linoleic acid and *trans* fatty acid composition of cow's milk fat produced in lowlands and highlands. *J Dairy Res*, **2001**, *68*, 519–523.

Craig-Schmidt, M.C. Consumption of *trans* fatty acids. In: Sébédio, J.-L.; Christie, W.W. (eds), *Trans* Fatty Acids in Human Nutrition, The Oily Press, Dundee, Scotland, 1998; pp. 59–114.

Cruz-Hernandez, C.; Z. Deng; J. Zhou; A.R. Hill; M.P. Yurawecz; P. Delmonte; M.M. Mossoba; M.E.R. Dugan; J.K.G. Kramer. Methods to analyze conjugated linoleic acids (CLA) and *trans*-18:1 isomers in dairy fats using a combination of GC, silver ion TLC-GC, and silver ion HPLC. *J AOAC Internat*, **2004**, *87*, 545–562.

Cruz-Hernandez, C.; J.K.G. Kramer; J. Kraft; V. Santercole; M. Or-Rashid; Z. Deng; M.E.R. Dugan; P. Delmonte; M.P. Yurawecz. Systematic analysis of *trans* and conjugated linoleic acids in the milk and meat of ruminants. In Yurawecz, M.P.; Kramer, J.K.G.; Gudmundsen, O.; Pariza, M.W.; Banni, S. (eds) Advances in Conjugated Linoleic Acid Research, Volume 3. AOCS Press, Champaign, IL, 2006; pp 45–93.

Cruz-Hernandez, C.; J.K.G. Kramer; J.J. Kennelly; B.M. Sorensen; E.K. Okine; L.A. Goonewardene; R.J. Weselake. Evaluating the CLA and *trans* 18:1 isomers in milk fat of dairy cows fed increasing amounts of sunflower oil. *J Dairy Sci*, **2007**, *90*, 3786–3801.

Dannenberger, D.; G. Nuernberg; N. Scollan; W. Schabbel; H. Steinhart; K. Ender; and K. Nuernberg. Effect of diet on the deposition of n-3 fatty acids, conjugated linoleic and C18:1*trans* fatty acid isomers in muscle lipids of German Holstein bulls. *J Agric Food Chem*, **2004**, *52*, 6607–6615.

Dannenberger, D.; K. Nuernberg; G. Nuernberg; N. Scollan; H. Steinhart; K. Ender.Effects of pasture vs. concentrate diet on CLA isomer distribution in different tissue lipids of beef cattle. *Lipids*, **2005**, *40*, 589–598.

Delmonte, P.; J.A.G. Roach; M.M. Mossoba; G. Losi; M.P. Yurawecz. Synthesis, isolation, and GC analysis of all the 6,8- to 13,15-*cis/trans* conjugated linoleic acid isomers. *Lipids*, **2004,** *39*, 185–191.

Delmonte, P.; A. Kataoka; B.A. Corl; D.E. Bauman; M.P. Yurawecz. Relative Retention order of all isomers of *cis/trans* conjugated linoleic acid FAME from the systematic analysis of *trans* and CLA 6,8- to 13,15-positions using silver ion HPLC with two elution systems. *Lipids*, **2005**, *40*, 509–514.

Department of Health and Human Services, FDA 21 CFR Part 101 [Docket No. 94P-0036] Food labeling: *Trans* fatty acids in nutrition labeling; nutrient content claims, and health claims; final rule, Federal Register 68, No. 133, July 11, 2003, pp 41434-41506

Destaillats, F.; P. Angers. Base-catalyzed derivatization methodology for FA analysis. Application to milk fat and celery seed lipid TAG. *Lipid*, **2002a**, *37*, 527–532.

Destaillats, F.; P. Angers. Evidence of [1,5] sigmatropic rearrangements of CLA in heated oils. *Lipid*, **2002b**, *37*, 435–438.

Destaillats, F.; P. Angers. Direct sequential synthesis of conjugated linoleic acid isomers from $\Delta^{7,9}$ to $\Delta^{12,14}$. *Eur J Lipid Sci Technol*, **2003**, *105*, 3–8.

Destaillats, F.; J.P. Trottier; J.M.G. Galvez; P. Angers. Analysis of α-linolenic acid biohydrogenation intermediates in milk fat with emphasis on conjugated linolenic acids. *J Dairy Sci*, **2005**, *88*, 3231–3239.

Destaillats, F.; P.-A. Golay; F. Joffre; M. de Wispelaere; B. Hug; F. Giuffrida; L. Fauconnot; F. Dionisi. Comparison of available analytical methods to measure *trans*-octadecenoic acid isomeric profile and content by gas-liquid chromatography in milk fat. *J Chromatogr A*, **2007**, *1145*, 222–228.

Dionisi, F.; Golay, P.A.; Fay, L.B. Influence of milk fat presence on the determination of trans fatty acids in fats used for infant formulae. *Anal Chim Acta*, **465**, *2002,* 395-407.

Duchateau, G.S.M.J.E.; H.J. van Osten; M.A. Vasconcellos. Analysis of *cis*- and *trans*-fatty acid isomers in hydrogenated and refined vegetable oils by capillary gas-liquid chromatography, *J Am Oil Chem Soc*, **1996**, *73*, 275–282.

Dugan, M.E.R.; J.K.G. Kramer; W.M. Robertson; W.J. Meadus; N. Aldai; D.C. Rolland. Comparing subcutaneous adipose tissue in beef and muskox with emphasis on *trans* 18:1 and conjugated linoleic acids. *Lipids*, **2007**, *42*, 509–518.

Eifert, E.C.; R.P. Lana; D.P.D. Lanna; W.M. Leopoldino; P.B. Arcuri; M.I. Leão; M.R. Cota; S.C. Valadares Filho. Milk fatty acid profile of cows fed monensin and soybean oil in early lactation. *Rev Bras Zootecn*, **2006**, *35*, 219–228.

El-Hamdy, A.H.; W.W. Christie.Preparation of methyl esters of fatty acids with trimethylsulphonium hydroxide -an appraisal *J Chromatography*, **1993**, *630*, 438–441.

Firestone, D.; P. LaBouliere. Determination of isolated *trans* isomers by infrared spectrophotometry, *J Assoc Off Anal Chem*, **1965**, *48*, 437–443.

Firestone, D.; A. Sheppard. Determination of *trans* fatty acids. In: Christie WW (ed), Advances in Lipid Methodology-One, The Oily Press, Ayr, UK, 1992; pp. 273–322.

Fritsche, J.; H. Steinhart. Analysis of *trans* fatty acids. In: Mossoba, M.M.; McDonald, R.E. (eds), New Trends in Lipid Analysis, AOCS Press, Champaign, IL, 1997; pp 234–256.

Fritsche, S.; T.S. Rumsey; M.P. Yurawecz; Y. Ku; J. Fritsche. Influence of growth promoting implants on fatty acid composition including conjugated linoleic acid isomers in beef fat. *Eur Food Res Technol*, **2001**, *212*, 621–629.

Golay, P.-A.; F. Dionisi; B. Hug; F. Giuffrida; F. Destaillats. Direct quantification of fatty acids in dairy powders with special emphasis on *trans* fatty acid content *Food Chem*, **2006**, *101*, 1115–1120.

Griinari, J.M.; D.E. Bauman. Biosynthesis of conjugated linoleic acid and its incorporation into meat and milk in ruminants. In: Yurawecz, M.P.; Mossoba, M.M.; Kramer, J.K.G.; Pariza, M.W.; Nelson, G.J. (eds) Advances in Conjugated Linoleic Acid Research Volume 1. AOCS Press, Champaign, IL, 1999; pp 180–200.

Griinari, J.M.; B.A. Corl; S.H. Lacy; P.Y. Chouinard; K.V.V. Nurmela; D.E. Bauman. Conjugated linoleic acid is synthesized endogenously in lactating dairy cows by Δ9 desaturase. *J Nutr*, **2000**, *130*, 2285–2291.

Grob, K. Split and Splitless Injection in Capillary GC, Wiley/VCH, Weinheim, Germany, 2001.

Grob, K.; M. Biedermann. The two options for sample evaporation in hot GC injectors: Thermospray and band formation. Optimization of conditions and injector design. *Anal Chem*, **2002**, *74*, 10–16.

Harrick, N.J. Internal Reflection Spectroscopy, Wiley-Interscience, New York, NY, 1967.

Henninger, M.; F. Ulberth. *Trans* fatty acid content of bovine milk fat. *Milchwissenschaft*, **1994**, *49*, 555–558.

Hristov, A.N.; L.R. Kennington; M.A. McGuire; C.W. Hunt. Effects of diets containing linoleic acid- or oleic acid-rich oils on ruminal fermentation and nutrient digestibility, and performance and fatty acid composition of adipose tissue and muscle tissues of finishing cattle. *J Anim Sci*, **2005**, *83*, 1312–1321.

Horrocks, L.A. Content, composition, and metabolism of mammalian and avian lipids that contain ether groups, In: Snyder F (ed), Ether Lipids, Chemistry and Biology, Academic Press, New York, 1972; pp 177–272.

House, S.D. Determination od total, saturated, monounsaturated fats in foodstuffs by hydrolytic extraction and gas chromatographic quantitation: Collaborative study. *J AOAC Internat*, **1997**, *90*, 555–563.

Ismail, A.A.; A. Nicodemo; J. Sedman; F.R. van de Voort; I.E. Holzbauer. Infrared spectroscopy of lipids: Principles and applications. In: Hamilton, R.J.; Cast, J. (eds), Spectral Properties of Lipids, Sheffield Academic Press, Sheffield, England / CRC Press, Boca Raton, FL, 1999; pp 235–269.

ISO 661, Animal and vegetable fats and oils—Preparation of test sample, Second edition 1989-06-15.

ISO/FDIS 5509, Animal and vegetable fats and oils—Preparation of methyl esters of fatty acids, 1999.

ISO 15304, Animal and vegetable fats and oils—Determination of the content of *trans* fatty acid isomers of vegetable fats and oils—Gas chromatographic method, (Corrected version 2003-05-15).

Jensen, R.G. The composition of bovine milk lipids: January 1995 to December 2000. *J Dairy Sci*, **2002**, *85*, 295–350.

JOCS 2.4.1.2-1996, Preparation of methyl esters of fatty acids (boron trifluoride-methanol method). Standard Methods for the Analysis of Fats and Oils and Related Materials, Japan Oil Chem. Soc. Ed.), 2.4.1.2-1996, 1996.

JOCS Kyo -4-1996, Standard Methods for the Analysis of Fats and Oils and Related Materials, Japan Oil Chem. Soc. ed.), Kyo-4-1996, pp. 1-12, 1996.

Juanéda, P.; S. Brac de la Pérrière; J.L. Sébédio; S. Grégoire. Influence of heat and refining on formation of CLA isomers in sunflower oil. *J Am Oil Chem Soc*, **2003**, *80*, 937–94.

Kraft, J.; M. Collomb; P. Möckel; R. Sieber; G. Jahreis. Differences in CLA isomer distribution of cow's milk lipids. *Lipids*, **2003**, *38*, 657–664.

Kraft, J.; J.K.G. Kramer; F. Schoene; J.R. Chambers; G. Jahreis. Extensive analysis of long-chain polyunsaturated fatty acids, CLA, *trans*-18:1 isomers, and plasmalogenic lipids in different retail beef types. *J Agric Food Chem*, **2008**, *56*, 4775–4782.

Kramer, J.K.G.; J. Zhou. Conjugated linoleic acids and octadecenoic acids: Extraction and isolation of lipids. *Eur J Lipid Sci Technol*, **2001**, *103*, 594–600.

Kramer, J.K.G.; N. Sehat; J. Fritsche; M.M. Mossoba; K. Eulitz; M.P. Yurawecz; Y. Ku. Separation of Conjugated Linoleic Acid Isomers. In: Yurawecz, M.P.; Mossoba, M.M.; Kramer, J.K.G.; Pariza, M.W.; Nelson, G.J. (eds), Advances in Conjugated Linoleic Acid Research, Volume 1, AOCS Press, Champaign, IL, USA, 1999; pp 83–109.

Kramer, J.K.G.; C.B. Blackadar; J. Zhou. Evaluation of two GC columns (60-m SUPELCO-WAX 10 and 100-m CP Sil 88) for analysis of milkfat with emphasis on CLA, 18:1, 18:2, and 18:3 isomers, and short- and long-chain FA. *Lipids*, **2002**, *37*, 823–835.

Kramer, J.K.G.; C. Cruz-Hernandez; J. Zhou. Conjugated linoleic acids and octadecenoic acids: Analysis by GC. *Eur J Lipid Sci Technol*, **2001**, *103*, 600–609.

Kramer, J.K.G.; V. Feller; M.E.R. Dugan; F.D. Sauer; M.M. Mossoba; M.P. Yurawecz. Evaluating acid and base catalysts in the methylation of milk and rumen fatty acids with special emphasis on conjugated dienes and total *trans* fatty acids. *Lipids*, **1997**, *32*, 1219–1228.

Kramer, J.K.G.; M. Hernandez; C. Cruz-Hernandez; J. Kraft; M.E.R. Dugan. Combining results of two GC separations partly achieves determination of all *cis* and *trans* 16:1, 18:1, 18:2, 18:3 and CLA isomers of milk fat as demonstrated using Ag-ion SPE fractionation. *Lipids*, **43**, *2008*, 259–273.

Larsson, S.C.; L. Bergkvist; A. Wolk. High-fat dairy food and conjugated linoleic acid intakes in relation to colorectal cancer incidence in the Swedish mammography cohort. *Am J Clin Nutr*, **2005**, *82*, 894–900.

Leheska, J.M.; L.D. Thompson; J.C. Howe; E. Hentges; J. Boyce; J.C. Brooks; B. Shriver; L. Hoover; M.F. Miller. Effects of conventional and grass-feeding systems on the nutrient composition of beef. *J Anim Sci*, **2008**, *86*, 3575–3585.

Lepage, G. and C.C. Roy. Direct transesterification of all classes of lipid in a one-step reaction, *J Lip Res*, **1986**, *27*, 114–120.

Li J.; Y. Fan; Z. Zhang; H. Yu; Y. An; J.K.G. Kramer; Z. Deng. Evaluating the *trans* fatty acid, CLA, PUFA and erucic acid diversity in human milk from five regions in China. *Lipids*, **2009**, *44*, 257–271.

Mahadevan, V.; C.V. Viswanathan; F. Phillips. Conversion of fatty acid aldehyde dimethyl acetals to the corresponding alk-1-enyl methyl ethers (substituted vinyl ethers) during gas-liquid chromatography. *J Lipid Res*, **1967**, *8*, 2–6.

Martin, J.-C., and K. Valeille. Conjugated linoleic acids: All the same or to everyone its own function? *Reprod Nutr Develop*, **2002**, *42*, 525–536.

Mendis, S.; Cruz-Hernandez; W.M.N. Ratnayake.Fatty Acid Profile of Canadian Dairy Products with Special Attention to the *trans*-Octadecenoic Acid and Conjugated Linoleic acid isomers. *J AOAC Internat*, **2008**, *91*, 811–819.

Milosevic, M; V. Milosovic; J.K.G. Kramer; H. Azizian; M.M. Mossoba. Determining low levels of *trans* fatty acids in foods by an improved ATR-FTIR procedure. *Lipid Technol*, **2004**, *16*, 229–231.

Mirabella, F.M. (ed) Internal Reflection Spectroscopy, Practical Spectroscopy Series, Vol 15, Marcel Dekker, New York, NY, 1992.

Mjøs, S.A. Identification of fatty acids in gas chromatography by application of different temperature and pressure programs on a single capillary column. *J Chromatogr A*, **2003**, *1015*, 151–161.

Molekentin, J.; D. Precht. Optimized analysis of *trans*-octadecenoic acids in edible fats. *Chromatographia*, **1995**, *41*, 267–272.

Mosley, E.E.; A.L. Wright; M.K. McGuire; M.A. McGuire. *trans* Fatty acids in milk produced by women in the United States. *Am J Clin Nutr*, **2005**, *82*, 1292–1297.

Mossoba, M.M.; M. Adam; T. Lee. Rapid determination of total *trans* fat content. An Attenuated Total Reflection infrared spectroscopy international collaborative study, *J AOAC Internat*, **2001a**, *84*, 1144–1150.

Mossoba, M.M.; J.K.G. Kramer; J. Fritsche; M.P. Yurawecz; K. Eulitz; Y. Ku; J.I. Rader. Application of standard addition to eliminate conjugated linoleic acid and other interferences in the determination of total *trans* fatty acids in selected food products by infrared spectroscopy. *J Am Oil Chem Soc*, **2001b**, *78*, 631–634.

Mossoba, M.M.; J.K.G. Kramer; P. Delmonte; M.P. Yurawecz; J.I. Rader. Official methods for the determination of *trans* fat. AOCS Press, Urbana, IL, 2003.

Mossoba, M.M.; J.K.G. Kramer; P. Delmonte; M.P. Yurawecz; J.I. Rader. Determination of *trans* fats by gas chromatography and infrared methods. In: Kodali, D.R.; List, G.R. (eds), *Trans* Fats Alternatives, AOCS Press, 2005; pp 47–70.

Mossoba, M.M.; J.K.G. Kramer; V. Milosevic; M. Milosevic; H. Azizian. Interference of saturated fats in the determination of low levels of *trans* fats (below 0.5%) by infrared spectroscopy. *J Am Oil Chem Soc*, **2007a**, *84*, 339–342.

Mossoba, M.M.; V. Milosevic; M. Milosevic; J.K.G. Kramer; and H. Azizian. Determination of total *trans* fats and oils by infrared spectroscopy for regulatory compliance. *Anal. Bioanal. Chem.*, **2007b**, *389*, 87–92.

Mossoba, M.M.; A. Seiler; J.K.G. Kramer; V. Milosevic; M. Milosevic; H. Azizian; H. Steinhart. Nutrition Labeling: Rapid Determination of Total *Trans* Fats by Using Internal Reflection Infrared Spectroscopy and a Second Derivative Procedure, *J Am Oil Chem Soc*, (submitted) **2009**.

Mossoba, M.M.; R.E. McDonald; J.A.G. Roach; D.D. Fingerhut; M.P. Yurawecz; N. Sehat. Spectral confirmation of *trans* monounsaturated C_{18} fatty acid positional isomers. *J Am Oil Chem Soc*, **1997**, *74*, 125–130.

Mossoba, M.M.; M.P. Yurawecz; R.E. McDonald. Rapid determination of the total *trans* content of neat hydrogenated oils by Attenuated Total Reflection spectroscopy, *J Am Oil Chem Soc*, **1996**, *73*, 1003-1009; and references therein.

Murrieta, C.M.; B.W. Hess; D.C. Rule. Comparison of acidic and alkaline catalysts for preparation of fatty acid methyl esters from ovine muscle with emphasis on conjugated linoleic acid. *Meat Sci*, **2003**, *65*, 523–529.

Nuernberg, K.; G. Nuernberg; K. Ender; S. Lorenz; K. Winkler; R. Rickert; H. Steinhart. N-3 fatty acids and conjugated linoleic acids of *longissimus* muscle in beef cattle. *Eur J Lipid Sci Technol*, **2002**, *104*, 463–471/

Nuernberg, K.; G. Nuernberg; K. Ender; D. Dannenberger; W. Schabbel; S. Grumbach; W. Zupp; H. Steinhart. Effect of grass *vs.* concentrate feeding on the fatty acid profile of different fat depots in lambs. *Eur J Lipid Sci Technol*, **2005**, *107*, 737–745.

Pariza, M.P.; Y. Park; M.E. Cook. The biologically active isomers of conjugated linoleic acid. *Progr Lipid Res*, **2001**, *40*, 283–298.

Park, Y.; K.J. Albright; Z.Y. Cai; M.W. Pariza. Comparison of methylation procedures for conjugated linoleic acid and artifact formation by commercial (trimethylsilyl)diazomethane. *J Agr Food Chem*, **2001**, *49*, 1158–1164.

Parodi, P.W. Conjugated linoleic acid in foods. In: Sébédio J-L, Christie WW, Adlof R (eds) Advances in Conjugated Linoleic Acid Research Volume 2. AOCS Press, Champaign, IL, 2003; pp 101–122.

Peene, J.J.; J. de Zeeuw; F. Biermans; L. Joziasse. CP- Select CB for FAME, A new highly polar bonded stationary phase with a temperature stability up to 290°C optimized for analyzing *cis-* and *trans* FAME isomers with GC, #P-147, Varian Inc., Middelburg, The Netherlands. 2004.

Piperova, L.S.; B.B. Teter; I. Bruckental; J. Sampugna; S.E. Mills; M.P. Yurawecz; J. Fritsche; Y. Ku; R.A. Erdman. Mammary lipogenic enzyme activity, *trans* fatty acids and conjugated linoleic acids are altered in lactating dairy cows fed a milk fat-depressing diet. *J Nutr*, **2000**, *130*, 2568–2574.

Precht, D.; J. Molkentin. Rapid analysis of the isomers of *trans*-octadecenoic acid in milk fat. *Int Dairy J*, **1996**, *6*, 791–809.

Precht, D.; J. Molkentin. Vergleich der Fettsäuren und der Isomerenverteilung der *trans*-C18:1-Fettsäuren von Milkfett, Margarine, Back-, Brat- und Diätfetten. Kieler Milchwirtschaftliche *Forschungsberichte*, **1997**, *49*, 17–34.

Precht, D.; J. Molkentin. C18:1, C18:2 and C18:3 *trans* and *cis* fatty acid isomers including conjugated *cis*-9,*trans*-11 linoleic acid (CLA) as well as total fat composition of German human milk lipids. *Nahrung*, **1999**, *43*, 233–244.

Precht, D.; J. Molketin. Identification and quantitation of *cis/trans* C16:1 and C17:1 fatty acid positional isomers in German human milk lipids by thin-layer chromatography and gas chromatography/mass spectrometry. *Eur J Lipid Sci Technol*, **2000a**, *102*, 102–113.

Precht, D.; J. Molkentin. Frequency distribution of conjugated linoleic acid and *trans* fatty acid contents in European bovine milk fats. *Milchwissenschaft*, **2000b**, *55*, 687–691.

Precht, D.; J. Molkentin. Recent trends in the fatty acid composition of German sunflower margarines, shortenings and cooking fats with emphasis on individual C16:1, C18:1, C18:2 C18:3 and C20:1 *trans* isomers. *Nahrung*, **2000c**, *44*, 222–228.

Precht D. and J. Molkentin. Overestimation of linoleic acid and *trans*-C18:2 isomers in milk fats with emphasis on *trans* Δ9, *trans* Δ12-octadecadienoic acid. *Milchwissenschaft*, **2003**, *58*, 30–34.

Rajakangas, J.; S. Basu; I. Salminen; M. Mutanen. Adenoma growth stimulation by the *trans*-10, *cis*-12 isomer of conjugated linoleic acid (CLA) is associated with changes in mucosal NF-B and cyclin D1 protein levels in the Min Mouse. *J Nutr*, **2003**, *133*, 1943–1948.

Ratnayake, W.M.N. Analysis of *trans* fatty acids. In: Sébédio J-L, Christie WW (eds), *Trans Fatty Acids in Human Nutrition*, The Oily Press, Dundee, Scotland, 1998; pp. 115–161.

Ratnayake, W.M.N. Analysis of dietary *trans* fatty acids. *J Oleo Sci*, **2001**, *50*, 73–86.

Ratnayake, W.M.N. Overview of methods for the determination of *trans* fatty acids by gas chromatography, silver-ion thin-layer chromatography, silver-ion liquid chromatography, and gas chromatography/mass spectrometry. *J AOAC Internat*, **2004**, *87*, 523–539.

Ratnayake, W.M.N.; C. Cruz-Hernandez. Analysis of trans fatty acids of partially hydrogenated vegetable oils and dairy products. In: Destaillats, F.; Sébédio, J.-L.; Dionisi, F.; Chardigny, J.-M. (eds), Trans Fatty Acids in Human Nutrition - Second Edition, The Oily Press, Bridgwater, UK, 2009; Chapter 5.

Ratnayake, M.N.; C. Zehaluk. *Trans* fatty acids in foods and their labeling regulations. In: Akoh CC, Lai O-M (eds), Healthful Lipids, AOCS Press, Champaign, IL, 2005; pp 1–32.

Ratnayake, W.M.N.; L.J. Plouffe; E. Pasquier; C. Gagnon. temperature-sensitive resolution of *cis*- and *trans* fatty acid isomers of partially hydrogenated vegetable oils on SP-2560 and CP-Sil 88 capillary columns. *J AOAC Internat*, **2002**, *85*, 1112–1118.

Ratnayake, W.M.N.; S.L. Hansen; M.P. Kennedy. Evaluation of the CP-Sil 88 and SP-2560 GC columns used in the recently approved AOCS official method Ce 1h-05: Determination of *cis*-, *trans*-, saturated, monounsaturated, and polyunsaturated fatty acids in vegetable or non-ruminant animal oils and fats by capillary GLC method. *J Am Oil Chem Soc*, **2006**, *83*, 475–488.

Ratnayake, W.M.N.; C. Gagnon; L. Dumais; W. Lillycrop; L. Wong; M. Meleta; P. Calway. *Trans* fatty acid content of Canadian margarines prior to mandatory *trans* fat labelling. *J Am Oil Chem Soc*, **2007**, *84*, 817–825.

Regulations Amending the Food and Drug Regulations (Nutrition Labelling, Nutrient Content Claims and Health Claims). Department of Health, Canada Gazette, Part 11. January 1, 2003. http://canadagazette.gc.ca/partII/2003/20030101/html/sor11-e.html

Reedy, G. and M.M. Mossoba. Matrix isolation GC-FTIR. In: Mossoba MM (ed), Spectral Methods in Food Analysis, Marcel Dekker Inc., New York, NY, 1999; pp 325–396.

Roach, J.A.G.; M.P. Yurawecz; J.K.G. Kramer; M.M. Mossoba; K. Eulitz; Y. Ku. Gas chromatography-high resolution selected-ion mass spectrometric identification of trace 21:0 and 20:2 fatty acids eluting with conjugated linoleic acid isomers. *Lipids*, **2000**, *35*, 797–802.

Roach, J.A.G.; M.M. Mossoba; M.P. Yurawecz; J.K.G. Kramer. Chromatographic separation and identification of conjugated linoleic acid isomers. *Anal Chim Acta*, **2002**, *465*, 207–226.

Roy, A.; A. Ferlay; K.J. Shingfield; Y. Chilliard. Examination of the persistency of milk fatty acid composition responses to plant oils in cows given different basal diets, with particular emphasis on *trans*-C18:1 fatty acids and isomers of conjugated linoleic acid. *Anim Sci*, **2006**, *82*, 479–492.

Roy, A.; J.M. Chardigny; D. Bauchart; A. Ferlay; S. Lorenz; D. Durand; D. Gruffat; Y. Faulconnier; J.-L. Sébédio; Y. Chilliard. Butters rich either in *trans*-10-C18:1 or in *trans*-11-C18:1 plus *cis*-9, *trans*-11 CLA differentially affect plasma lipids and aortic fatty streak in experimental atherosclerosis in rabbits. *Animal*, **2007**, *1*, 467–476.

Rozema, B.; B. Mitchell; D. Winters; A. Kohn; D. Sullivan; E. Meinholz. Proposed Modifications to AOAC 996.06, Optimizing the Determination of *trans* Fatty Acids: Presentation of Data. *J AOAC Internat*, **2008**, *91*, 92–97.

Santercole, V.; R. Mazzette; E.P.L. De Santis; S. Banni; L. Goonewardene; J.K.G. Kramer. Total lipids of Sarda sheep meat that include the fatty acid and alkenyl composition and the CLA and *trans*-18:1 isomers. *Lipids*, **2007**, *42*, 361–382.

Satchithanandam, S.; C.J. Oles; C.J. Spease; M.M. Brandt; M.P. Yurawecz; J.I. Rader. *Trans*, saturated, and unsaturated fat in foods in the United States prior to mandatory *trans*-fat labeling, *Lipids*, **2004**, *39*, 11–18.

Sébédio, J.L.; M. Catte; M.A. Boudier; J. Prevost; A. Grandgirard. Formation of fatty acid geometric isomers and cyclic fatty acid monomers during the finish frying of frozen pre-fried potatoes. *Food Res Int*, **1996**, *29*, 109–116.

Sehat, N.; R.Rickert; M.M. Mossoba; J.K.G. Kramer; M.P. Yurawecz; J.A.G. Roach; R.O. Adlof; K.M. Morehouse; J. Fritsche; K. Eulitz; H. Steinhart; Y. Ku. Improved separation of conjugated linoleic acid methyl esters by silver-ion high-performance liquid chromatography. *Lipids*, **1999**, *34*, 407–413.

Shantha, N.C.; E.A. Decker; Z. Ustunol. Conjugated linoleic acid concentration in processed cheese. *J Am Oil Chem Soc*, **1992**, *69*, 425–428.

Shantha, N.C.; L.N. Ram; J. O'Leary; C.L. Hicks; E.A. Decker. Conjugated linoleic acid concentration in dairy products as affected by processing and storage. *J Food Sci*, **1995**, *60*, 695–697, 720.

Shingfield, K.J.; C.K. Reynolds; B. Lupoli; V. Toivonen; M.P. Yurawecz; P. Delmonte; J.M. Griinari; A.S. Grandison; D.E. Beever. Effect of forage type and proportion of concentrate in the diet on milk fatty acid composition in cows given sunflower oil and fish oil. *Anim Sci*, **2005**, *80*, 225–238.

Shingfield, K.J.; C.K. Reynolds; G. Hervás; J.M. Griinari; A.S. Grandison; D.E. Beever. Examination of the persistency of milk fatty acid composition responses to fish oil and sunflower oil in the diet of dairy cows. *J Dairy Sci*, **2006**, *89*, 714–732.

Shirasawa, S.; M. Shiota; H. Arkawa; Y. Shigematsu; K. Yokomizo; T. Shionoya; T. Okanoto; Y. Miyazaki; S. Takahashi; K. Himata. Quantitative determination of *trans*-fatty acids in oils and fats by capillary gas chromatography: Results of a JOCS collaborative study. *J Oleo Sci*, **2007**, *56*, 405–415.

Stein, R.A. and V. Slawson. A model for fatty aldehyde dimethyl acetal gas-liquid chromatography. The conversion of octadecanal dimethyl acetal to methyl 1-octadecenyl ether. *J Chromatogr*, **1966**, *25*, 204–212.

Stoffel, W.; F. Chu; E.H. Ahrens. Analysis of long-chain fatty acids by gas-liquid chromatography. Micromethod for preparation of methyl esters. *Anal Chem*, **1959**, *31*, 307–308.

Supelco Technical Report Discovery7 Ag-Ion SPE for FAME fractionation and *cis/trans* separation. Sigma-Aldrich.Com/Supelco, email: supelco@sial.com, Bellefonte, PA, 2006.

Tricon, S.; G.C. Burdge; S. Kew; T. Banerjee; J.J. Russell; E.L. Jones; R.F. Grimble; C.M. Williams; P. Yaqoob; P.C. Calder. Opposing effects of *cis*-9,*trans*-11 and *trans*-10,*cis*-12 conjugated linoleic acid on blood lipids in healthy humans. *Am J Clin Nutr*, **2004**, *80*, 614–620.

Ulberth, F.; R.G. Gabernig; F. Schrammel. Flame-ionization detector response to methyl, ethyl, propyl, and butyl esters of fatty acids. *J Am Oil Chem Soc*, **1999**, *76*, 263–266.

Venkata Rao, P.; S. Ramachandran; D.G. Cornwell. Synthesis of fatty acid aldehydes and their cyclic acetals (new derivatives for the analysis of plasmalogens). *J Lipid Res*, **1967**, *8*, 380–390.

Vickers, A.K.; M. Hastings; R. Lautamo; R. Davis; S. Watkins. New High polarity bis (cyanopropyl) siloxane stationary phase for GC resolution of positional and geometric isomers of fatty acid methyl esters. Agilent Technologies, November 11, 2004, 5989-1817EN

Wallace, R.J.; N. McKain; K.J. Shingfield; E. Devillard. Isomers of conjugated linoleic acids are synthesized via different mechanisms in ruminal digesta and bacteria. *J Lipid Res*, **2007**, *48*, 2247–2254

Wsowska, I.; M.R.G. Maia; K.M. Niedïwiedzka; M. Czauderna; J.M.C. Ramalho Ribeiro; E. Devillard; K.J. Shingfield; R.J. Wallace. Influence of fish oil on ruminal biohydrogenation of C18 unsaturated fatty acids. *Br J Nutr*, **2006**, *95*, 1199–1211.

Werner, S.A.; L.O. Luedecke; T.D. Shultz. Determination of conjugated linoleic acid content and isomer distribution in three cheddar-type cheeses: Effects of cheese cultures, processing, and aging. *J Agric Food Chem*, **1992**, *40*, 1817–1821.

Winterfeld, M.; H. Debuch. Die Lipoide einiger Gewebe und Organe des Menschen. *Hoppe Seylers Z Physiol Chem*, **1966**, *345*, 11–21.

Wolff, R.L. *trans*-Polyunsaturated fatty acids in French edible rapeseed and soybean oils. *J Am Oil Chem Soc*, **1992**, *69*, 106–110.

Wolff, R.L. Analysis of alpha-linolenic acid geometrical isomers in deodorized oils by capillary gas-liquid chromatography on cyanoalkyl polysiloxane phases: A note of caution. *J Am Oil Chem Soc*, **1994**, *71*, 907–909.

Wolff, R.L. Content and distribution of *trans*-18:1 acids in ruminant milk and meat fats.

Their importance in European diets and their effects on human milk. *J Am Oil Chem Soc*, **1995**, *72*, 259–272.

Wolff, R.L. Characterization of *trans*-monounsaturated alkenyl chains in total plasmalogens (1-O-alk-1 -enyl-2-acyl glycerophospholipids) from sheep heart. *Lipids*, **2002**, *37*, 811–816.

Wolff, R.L.; C.C. Bayard. Improvement in the resolution of individual *trans*-18:1 isomers by capillary gas-liquid chromatography: use of a 100-m CP Sil 88 column. *J Am Oil Chem Soc*, **1995**, *72*, 1197–1201.

Wolff, R.L. and R.J. Fabien. Utilisation de l'isopropanol pour l'extraction de la matière grasse de produits laitiers et pour l'esterification subséquente des gras. *Le Lait*, **1989**, *69*, 33–46.

Wolff, R. L.; C. C. Bayard; R. J. Fabien. Evaluation of sequential methods for the determination of butterfat fatty acid composition with emphasis on trans-18:1 acids. Application to the study of seasonal variations in French butters. *J. Am. Oil Chem. Soc.* **1995**, *72*, 1471–1483.

Wolff, R.L. and D. Precht. A critique of 50-m CP-Sil 88 capillary columns used alone to assess *trans*-unsaturated FA in foods: The case of TRANSFAIR study. *Lipids*, **2002**, *37*, 627–629.

Wolff, R.L.; N.A. Combe; F. Destaillats; C. Boué; D. Precht; J. Molkentin; B. Entressangles. Follow-up of the Δ4 to Δ16 *trans*-18:1 isomer profile and content in French processed foods containing partially hydrogenated vegetable oils during the period 1995-1999. Analytical and nutritional implications. *Lipids*, **2000**, *35*, 815–825.

Wolff, R.L.; D. Precht; J. Molkentin. Occurrence and distribution profiles of *trans*-18:1 acids in edible fats of natural origin. In: Sébédio J-L, Christie WW (eds), *Trans* Fatty Acids in Human Nutrition, The Oily Press, Dundee, Scotland, 1998; pp 1–34.

Yurawecz, M.P.; J.K. Hood; J.A.G. Roach; M.M. Mossoba; D.H. Daniels; Y. Ku; M.W. Pariza; S.F. Chin. Conversion of allylic hydroxy oleate to conjugated linoleic acid and methoxy oleate by acid-catalyzed methylation procedures. *J Am Oil Chem Soc*, **1994**, *71*, 1149–1155.

Yurawecz, M.P.; J.K.G. Kramer; Y. Ku. Methylation procedure for conjugated linoleic acid. In: Yurawecz MP, Mossoba MM, Kramer JKG, Pariza MW, Nelson GJ (eds), Advances in Conjugated Linoleic Acid Research, Volume 1, AOCS Press, Champaign, IL, USA, 1999; pp 64–82.

Yurawecz, M.P.; J.A.G. Roach; N. Sehat; M.M. Mossoba; J.K.G. Kramer; J. Fritsche; H. Steinhart; Y. Ku. A new conjugated linoleic acid isomer, 7 *trans*, 9 *cis*-octadecadienoic acid, in cow milk, cheese, beef and human milk, and adipose tissue. *Lipids*, **1998**, *33*, 803–809.